Ordovician rhynchonelliformean brachiopods from Co. Waterford, SE Ireland: Palaeobiogeography of the Leinster Terrane

by

Maria Liljeroth, David A. T. Harper, Hilary Carlisle and Arne T. Nielsen

Acknowledgements
Financial support for the publication of this issue of Fossils and Strata
was provided by the Lethaia Foundation.

Contents

Ordovician rhynchonelliformean brachiopods from Co. Waterford, SE Ireland: palaeobiogeography of the Leinster Terrane

MARIA LILJEROTH, DAVID A.T. HARPER, HILARY CARLISLE AND ARNE T. NIELSEN

FOSSILS AND STRATA

THE LETHAIA FOUNDATION

Liljeroth, M., Harper, D.A.T., Carlisle, H. & Nielsen, A.T. 2017: Ordovician rhynchonelliformean brachiopods from Co. Waterford, SE Ireland: palaeobiogeography of the Leinster Terrane. Fossils and Strata, No. 62, pp. 1–164. doi: 10.111/let.12205.

The Irish Leinster Terrane, Co. Waterford, SE Ireland, has traditionally been associated with a palaeogeographical position close to East Avalonia, but a number of recent studies on, for example, zircon geochronology, geochemistry, isotope analyses and tectono- and lithostratigraphy strongly point to a Ganderian affinity. Together with the Monian Composite, Bellewstown, Grangegeeth, Central Newfoundland and Miramichi terranes, the Leinster Terrane formed part of the leading edge of Ganderia in association with the Popelogan-Victoria arc several hundred kilometres north of Avalonia during the Ordovician. The present faunal data together with the Ganderian studies and compiled palaeomagnetic data from other authors form the basis of new palaeobiogeographical reconstructions of the mid Darriwilian and early Sandbian palaeoplate configurations in the south Iapetus Ocean. The generic composition of the Tramore and Dunabrattin Limestone formations contains only few endemic forms and is dominated by a mix of taxa that originated in other parts of the world during the Early–early Late Ordovician including low-latitude Laurentia, high-latitude peri-Gondwana and mid-latitude Baltica. All the Ganderian terranes and East Avalonia contained mixed faunas during the mid Darriwilian – early Sandbian indicating that these sites together occupied a mid-latitudinal position comprising settling grounds for taxa from the surrounding, widely dispersed high-, low- and mid-latitude palaeoplates. The Tramore and Dunabrattin limestones furthermore contain brachiopod species that migrated to the Leinster Terrane from East Avalonia, Baltica, Laurentia and peri-Laurentia during the late Darriwilian – early Sandbian confirming the Ganderian terrane's close relationship with the other Iapetus-bordering sites. Oceanic wind-driven surface currents were the primary factor determining brachiopod dispersal in the Darriwilian, when the majority of taxa were associated with shelfal biofacies. The brachiopod migrational patterns reported for the Darriwilian in the present investigation together with the Mid Ordovician synthetic ocean surface circulation models form the basis of revised models of oceanic surface current flow for the early and late Darriwilian, respectively. The *gracilis* drowning diminished brachiopod provinciality in the early Sandbian, and the number of reported brachiopod provinces was reduced from five to four. The mid-latitude Celtic and Baltic provinces were thus obliterated and replaced by the combined Anglo-Welsh–Baltic Province. A total of 20 rhynchonelliformean brachiopod genera and 22 species are described from the upper Darriwilian – lower Sandbian Tramore and Dunabrattin Limestone formations (Leinster Terrane). Seven of these are new species and include *Atelelasma longisulcum* n. sp., *Colaptomena auduni* n. sp., *Dactylogonia costellata* n. sp., *Hesperorthis leinsterensis* n. sp., *Howellites hibernicus* n. sp., *Isophragma parallelum* n. sp. and *Platystrophia tramorensis* n. sp. as well as one new genus *Hibernobonites* n. gen. and the earliest recorded occurrence of a *Tetraphalerella*-like genus in the lowermost part of the Tramore Limestone. □ *Avalonia, biogeography, brachiopods, Darriwilian, Ganderia, Leinster Terrane, oceanic surface currents, Sandbian.*

Maria Liljeroth [foramser@gmail.com], DK-7950 Erslev, Mors, North Jutland, Denmark; David A.T. Harper [david.harper@durham.ac.uk], Palaeoecosystems Group, Department of Earth Sciences, Durham University, Durham DH1 3LE, UK; Hilary Carlisle [hcarlisle.belfast@gmail.com], Osborne Park Belfast BT9 6JN, UK; Arne T. Nielsen [arnet@ign.ku.dk], Department of Geosciences and Natural Resource Management, Øster Voldgade 10 DK-1350 Copenhagen, Denmark; manuscript received on 6/03/2016; manuscript accepted on 12/12/2016.

Introduction

The current work comprises a description of the rhynchonelliformean brachiopods from the upper Darriwilian – lower Sandbian Tramore Limestone Formation and lower Sandbian Dunabrattin Limestone Formation, Tramore area, Co. Waterford, southeast Ireland. The stratigraphy of the formations is updated with new lithological and faunal correlations between the formation units as well as biostratigraphical correlation with the international and British stages of Cooper & Sadler (2012) and international time-slices of Webby *et al.* (2004). New interpretations regarding the depositional settings and palaeoecology of the formations are likewise provided, including a new lithologically based model for palaeobathymetric characterization.

The Tramore fauna is unique in the sense that it contains a mixed fauna of, for example, Laurentian, Baltic and high-latitude peri-Gondwanan origins with no unequivocal relation to any of the defined biogeographical provinces in the Darriwilian. The complex faunal composition of the Darriwilian – Sandbian Leinster Terrane has led some authors to suggest a more oceanwards position for the terrane than traditionally applied, that is within the leading edge of East Avalonia (Owen & Parkes 2000) or a separate drifting/docking history (Fortey & Cocks 2003). The current work provides a detailed assessment of the brachiopod provinces described by Harper *et al.* (2013) and the faunas related to the investigated palaeoplates with focus on the Iapetus and Rheic oceans and the Tornquist Sea not only to resolve the source of the mixed faunal signature of the Leinster Terrane but also to reveal important details about the individual palaeoplates within the provinces that have hitherto not been assessed by the traditionally more general, global biogeographical analyses. The two time-slices applied herein mainly cover the early mid–latest Darriwilian and the early Sandbian, respectively, and largely include stratigraphy with a significant biostratigraphical overlap with the Tramore and Dunabrattin Limestone formations. This is an attempt to study high-resolution brachiopod biogeography on a wider geographical scale within very limited time intervals. Previous procedure for these investigations within the Darriwilian and Sandbian was often to combine the Dapingian and Darriwilian into one time-slice and to include the total Sandbian interval due to poor biostratigraphical constraint (see, e.g., Harper *et al.* 2013). By implication not all stratigraphical sections overlapping with the late Darriwilian or early Sandbian are included in the multivariate biogeographical analyses but the taxonomical compositions of the individual sites and provinces within the time-slices are compared with later and earlier faunas to give a more complete picture.

The Leinster Terrane has traditionally been illustrated as having close geographical proximity to East Avalonia based on faunal affinities and the presence of arc-related volcanic rocks (e.g. Parkes & Harper

1996; Cocks *et al.* 1997; Harper & Mac Niocaill 2002) but several analyses including detrital zircon geochronology, geochemistry, isotopic signatures, litho- and tectonostratigraphy and palaeomagnetism (e.g. van Staal *et al.* 1996, 1998, 2009, 2012; Collins & Buchan 2004; Valverde-Vaquero *et al.* 2006; Hibbard *et al.* 2007; Fyffe *et al.* 2009; Zagorevski *et al.* 2010; Pothier *et al.* 2015; Waldron *et al.* 2014) now point to a position within the leading edge of Ganderia close to the Popelogan-Victoria volcanic arc together with the Monian Composite (i.e. Rosslare and Anglesey), Bellewstown, Grangegeeth, Central Newfoundland and Miramichi terranes. Based on these results as well as compiled palaeomagnetic data from several authors (i.e. Torsvik *et al.* 1990, 1992, 1996; van der Voo *et al.* 1991; Trench & Torsvik 1992), the current study suggests new palaeobiogeographical reconstructions for the Iapetus Ocean for the mid Darriwilian and early Sandbian, respectively.

The Tramore fauna was not the only fauna of mixed origins in the late Darriwilian. In fact, East Avalonia and the Ganderian terranes contained varying numbers of biogeographically important taxa that originated in different parts of the world including Laurentia, peri-Laurentia and peri-Gondwana with an additional significant Baltic component present in the Ganderian terranes not reported from East Avalonia. This poses an interesting problem relating to geographical position and oceanic surface current systems that, when seen in context, aids explanation of the faunal affinities of the Leinster Terrane.

The studied time-slices were part of the Great Ordovician Biodiversification Event (GOBE) when global marine diversity increased dramatically and the rhynchonelliform brachiopods experienced major diversification on family, genus and species level. Orthide and strophomenide origination rates exceeded extinction rates across the Darriwilian – Sandbian transition on a global scale (Harper *et al.* 2004), although the present high-resolution study identifies sites in which diversity declined across the transition due to accelerated extinction and localized disappearance of specialized taxa.

Previous work

The continuous succession of volcanic and sedimentary deposits now assigned to the Duncannon Group was first described by Reed (1899) from a locality near Tramore town. He drew several sketches along the foreshore showing the relation between the formations of the group including the Tramore

Limestone Formation. Subsequent stratigraphical accounts by Mitchell *et al.* (1972), Williams (1976), Carlisle (1979), Harper & Parkes (2000), Wyse Jackson *et al.* (2001) and Key *et al.* (2005) and this study were based on the same area.

The brachiopod faunas of the Duncannon Group in County Waterford were described by Carlisle *in* Mitchell *et al.* (1972) and Carlisle (1979). The 1972 paper gave a preliminary palaeontological and stratigraphical account of the Duncannon Group formations, and the 1979 paper summarizes Carlisle's main work with a reinterpretation of the Tramore area faunas. Parkes (1994) published a monographic work on the brachiopod systematics from selected formations of the Duncannon Group primarily using Carlisle's original material collected in the Tramore area in the 1970s excluding the Tramore and Dunabrattin Limestone formations. Other palaeontological work includes descriptions of the bryozoans of the Tramore and Dunabrattin limestones and their palaeoenvironmental affinities (Wyse Jackson *et al.* 2001), a monographic work on the trilobites of the Duncannon Group published by Owen & Parkes (2000) and a crinoid described from the Tramore Limestone by Donovan (1985).

The early work by Williams (1976) recognized a division of Ireland based on the different faunal affinities. He divided the country into three distinct southwest–northeast oriented belts largely corresponding to (1) the Irish part of the Midland Valley Terrane (possibly together with the Grampian Terrane) of Scoto-Appalachian faunal affinity; (2) the Irish part of the Southern Uplands Terrane with a Southern Uplands faunal affinity; and (3) the Leinster Terrane with Anglo-Welsh and Baltic affinities. Palaeobiogeographical analyses of the Irish terranes were provided by, for example, Harper & Parkes (1989), Murphy *et al.* (1991), Harper (1992), Parkes (1992), Owen & Parkes (2000) and Key *et al.* (2005). Harper & Parkes (2000) recognized at least seven tectonic terranes in Ireland based on faunal evidence and the present study follows this division although with terrane names from Waldron *et al.* (2014). An overview of the palaeontological characteristics of the Irish terranes was provided by Harper & Parkes (1989).

The tectonic evolution of the Iapetus Ocean was described in detail by van Staal *et al.* (1996, 1998, 2009, 2012), Zagorevski *et al.* (2010), Pollock *et al.* (2012) and Pothier *et al.* (2015). Detailed accounts of the tectonic evolution of Avalonia and Ganderia were given by van Staal *et al.* (1996, 1998, 2009), Hibbard *et al.* (2007), Pollock *et al.* (2012) and Waldron *et al.* (2014).

Geological setting

The Tramore and Dunabrattin Limestone formations belong to the upper Darriwilian – lower Katian Duncannon Group of southeastern counties Waterford and Wexford, Ireland. The Duncannon Group comprises a suite of shallow- to deep-water impure carbonates and calcareous mudrocks overlain by basaltic, andesitic and rhyolitic arc volcanics accumulated on the Leinster Terrane (Carlisle 1979; Harper & Parkes 2000; Key *et al.* 2005).

Southeastern Ireland including the Leinster Terrane and the associated Leinster Basin formed part of the southern margin of the Iapetus Ocean through the Ordovician (e.g. Stillman 1978; Neuman & Harper 1992; Harper *et al.* 1996; Mac Niocaill *et al.* 1997; van Staal *et al.* 1998, 2012; Harper & Mac Niocaill 2002; Cocks & Torsvik 2011, 2013). Its Middle–Upper Ordovician volcanic rocks have previously been associated with subduction related arc–backarc volcanism by the leading edge, that is the northern margin, of the Avalonian microcontinent (e.g. Phillips *et al.* 1976; Stillman 1986; Parkes 1992; Tietzsch-Tyler & Sleeman 1994; Sleeman & McConnell 1995; Parkes & Harper 1996; Cocks *et al.* 1997; Mac Niocaill *et al.* 1997; Harper & Mac Niocaill 2002; Key *et al.* 2005; Brenchley & Rawson 2006; Brenchley *et al.* 2006; Harper *et al.* 2013) but recent results by other workers (see below) suggest that the volcanic rocks of the Leinster Terrane were more likely related to arc–backarc volcanism in the Popelogan-Victoria arc–Tetagouche-Exploits backarc basin (PVA-TEB) by the leading edge of the Ganderian microcontinent north of Avalonia. Owen & Parkes (2000) stressed that the Leinster Terrane had a more complex tectonic and biogeographical relationship with the Anglo-Welsh area than previously considered (i.e. by workers placing the terrane close to the northern East Avalonian margin) and suggested a more oceanwards position of this terrane. Fortey & Cocks (2003) suggested it had a separate drifting/docking history to East Avalonia based on brachiopod and trilobite faunas, although they illustrated the terrane as positioned on the northern East Avalonian margin.

Tectonomagmatic analyses of West Avalonian and west Ganderian basement rocks in the Appalachian orogen by Barr & White (1996) showed that Avalonia and Ganderia have different basement composition. Several other authors, applying one or more of the following types of data and methods including detrital zircon geochronology, geochemistry, isotopic signatures, litho- and tectonostratigraphy and palaeomagnetism (e.g. van Staal *et al.* 1996, 1998, 2009, 2012; Collins & Buchan 2004; Valverde-Vaquero *et al.* 2006; Hibbard *et al.* 2007; Fyffe *et al.* 2009; Zagorevski *et al.* 2010; Pothier *et al.* 2015; Waldron *et al.* 2014), identified Ganderia and its associated islands and volcanic arc terranes and/or the mutual palaeogeographical relationship between a selection of these and other terranes and micro-continents. Pollock *et al.* (2012) reviewed the palaeogeographical development of the peri-Gond-wanan realm of the Appalachian orogeny including the history of Ganderia and Avalonia and summa-rized their definitions.

The regional distribution of Ganderia and Aval-onia in Ireland and Great Britain is shown by Pol-lock *et al.* (2012, fig. 4). Terranes included in Ganderia are the Bellewstown (east Ireland), Gran-gegeeth (east Ireland) and Leinster–Lakesman (south Ireland–north England) terranes and the Monian Composite Terrane (Anglesey of northwest Wales and Rosslare of southeast Ireland). The Southern Uplands Terrane was also included in Ganderia by Pollock *et al.* (2012), but zircon geochronology studies by Waldron *et al.* (2014) strongly suggested a Laurentian affinity for the ter-rane. The regional distributions of the northern Appalachian terranes associated with Ganderia are illustrated in Pollock *et al.* (2012, fig. 3) and include the Central Newfoundland and Miramichi (New Brunswick, Maine) terranes. Following the results of van Staal *et al.* (1996, 1998, 2009, 2012), these terranes formed part of the Popelogan-Vic-toria arc–Tetagouche-Exploits backarc basin system created by subduction at the leading edge of Gan-deria during the progressive closure of the Iapetus Ocean. Harper & Mac Niocaill (2002) associated the Central Newfoundland and Miramichi terranes of Ganderia with the western part of the Popelo-gan-Victoria arc north of Avalonia but they still included the Leinster and Monian Composite ter-ranes as part of the East Avalonian margin. The Welsh Basin of East Avalonia includes Tremado-cian arc volcanics but the development of the Popelogan-Victoria arc–Tetagouche-Exploits back-arc basin system in the Floian (ca. 475 Ma; van Staal *et al.* 2012) changed both location and chem-istry of the volcanism and transformed the Welsh Basin into a backarc basin (Brenchley & Rawson 2006; Brenchley *et al.* 2006). The Leinster Terrane and Basin includes Sandbian acidic volcanics typi-cal of arc–backarc volcanism and was located oceanwards of the Welsh and Anglian basins (Brenchley & Rawson 2006), possibly in close proximity to the Popelogan-Victoria arc.

Tectonic history of Ganderia and Avalonia

Ganderia and Avalonia are peri-Gondwanan micro-continents that were situated along the northern margin of West Gondwana (Amazonia) and close to or juxtaposed against the West African margin, respectively, in the late Neoproterozoic (Fyffe *et al.* 2009; van Staal *et al.* 2009, 2012; Pothier *et al.* 2015; Waldron *et al.* 2014). Avalonia was situated at ca. 65°S during the mid–late Cambrian with Ganderia located north of this, and with a present-day strike length of at least 2500 km; Ganderia may have extended for over 23° of latitude allowing for a pos-sible along-strike connection with Avalonia (van Staal *et al.* 2012). The presence of the Cambrian trilobites *Kootenia* and *Baliella* in a limestone block of the Dunnage Mélange, central Newfoundland, shows that Ganderia formed at high latitudes along the Gondwanan margin (Pollock *et al.* 2012).

Ganderia started to separate from Amazonia around 505 Ma, opening the northern arm of the Rheic Ocean. The Rheic Ocean may initially have opened as a backarc basin with the Penobscot vol-canic arc positioned along the leading edge of Gan-deria from 584 to 515 Ma (Zagorevski *et al.* 2010; van Staal *et al.* 2012). Subsequent dextral-oblique motion moved most of Avalonia to a position just south of Ganderia, along the Amazonian margin and created a narrow intervening seaway of trapped oceanic lithosphere between the two microconti-nents (van Staal *et al.* 2012). Avalonia rifted from Gondwana in the late Tremadoc (479 Ma) (Murphy *et al.* 2004) opening the southern arm of the Rheic Ocean (van Staal *et al.* 2012). Comparable faunal assemblages support Gondwanan linkages with Aval-onia until the Tremadocian (Cocks *et al.* 1997; For-tey & Cocks 2003; Harper *et al.* 2013) after which the faunal provinciality indicates changing affinities from Gondwanan to Laurentian (Harper *et al.* 2013).

The Popelogan-Victoria arc was initiated around 475 Ma when subduction commenced again by the leading edge of Ganderia. The volcanic arc existed until the collision with Laurentia in the late Sand-bian (van Staal *et al.* 2012) or early Katian. It may have extended from beyond the eastern tip of Gan-deria along its complete east–west strike length, con-tinuing into the Famatina arc to the west along the present-day western margin of South America (see van Staal *et al.* 2012). Subduction of Avalonia beneath Ganderia took place during the Silurian–early Devonian after Ganderia had accreted to com-posite Laurentia.

Calculated drift rates of ca. 9 cm/a for the leading edge of Ganderia (500–455 Ma) and for Avalonia (490–460 Ma) indicate that they were situated on the same microplate after they had parted from Gondwana (van Staal *et al.* 2012) and they subsequently moved northeastwards by sinistral-oblique movement towards Laurentia through the Early–Late Ordovician (Pollock *et al.* 2012). The latitudinal drift rate for Avalonia decreased to 5 cm/a by ca. 460 Ma, which van Staal *et al.* (2012) considered to be a proxy for the drift rate of the trailing edge of Ganderia at this time. The ca. 4 cm/a discrepancy in drift rates between the Popelogan-Victoria arc and the trailing Ganderia margin is attributed to rifting in the intervening Tetagouche-Exploits backarc basin, which initiated at ca. 475 Ma when drift rates of Ganderia and Avalonia slowed down (van Staal *et al.* 2009). The rate discrepancy suggests a half-spreading rate of ca. 2 cm/a in the Tetagouche-Exploits basin producing a 600- to 800-km-wide ocean prior to 455 Ma (van Staal *et al.* 2012).

Pothier *et al.* (2015) presented an alternative tectonic model for Ganderia and Avalonia. In their mid Cambrian (ca. 500 Ma) and late Tremadocian (ca. 480 Ma) reconstructions, Ganderia and Avalonia are rotated more than 90° counterclockwise with the present (palaeo)northern margin of Ganderia juxtaposed against Amazonia and the (palaeo)southern margin of Avalonia facing eastwards towards an open ocean with Baltica in the distance. Pothier *et al.* (2015) suggested that after Avalonia had moved to a position just north of Ganderia by sinistral transpression in the late Tremadocian, Ganderia and Avalonia (incl. Meguma in which they included North Wales and Nova Scotia) moved northwestwards together along the Amazonian margin to leave Amazonia at a latitude of about 30°S while rotating clockwise to face the Laurentian margin. Their model was based on sediment provenance and a possible sinistral movement of Avalonia north of Ganderia the latter implying the necessity of placing these microcontinents in an 'upside down' orientation relative to the Gondwanan margin. The faunal analyses of this study do not readily support their model, however, as the Darriwilian Ganderian sites would be expected to include less Laurentian and Baltic brachiopod genera and a higher proportion of Gondwanan taxa than reported. The palaeomagnetically inferred latitudinal positions of central Newfoundland and the Welsh Basin through the Early–Late Ordovician, described below, do not support this either. The reconstruction of van Staal *et al.* (2012) does not contradict the influx of West African and Amazonian sediment to Meguma and Ganderia, respectively, in the mid Cambrian and late

Tremadocian as inferred by Pothier *et al.* (2015) as they placed Meguma and Avalonia very close to the West African margin, and juxtaposed Ganderia against Amazonia during these times.

The Iapetus Ocean

The Iapetus Ocean existed for about 170 millon years (ca. 580–420 Ma) and may have been as wide as 3000–5000 km estimated by faunal evidence and palaeomagnetic records (e.g. Cocks & Fortey 1982; Fortey & Cocks 1988; Harper 1992; Trench & Torsvik 1992; Harper *et al.* 1996). Palaeomagnetic analyses place the Early Ordovician Scottish segment of the Laurentian margin in low southerly latitudes of about 15–20°S (Torsvik *et al.* 1990, 1996) and East Avalonia about 60°S ±5–10°, the same latitude as the West African margin (Channel *et al.* 1992; Trench *et al.* 1992). The Miramichi Terrane may have been positioned at ca. 53°S in the early–mid Darriwilian although the age constraint on the Miramichi Group from which this palaeolatitude was obtained is not very precise (see McNicoll *et al.* 2002). During the Floian, Avalonia had moved to a position of about 41°S ±8° (Pollock *et al.* 2012) and in the latest Floian (470 Ma) palaeomagnetic analyses by Van der Voo *et al.* (1991) indicate that some Central Newfoundland Terrane localities, including Summerford, were at a latitudinal position of 31°S ±8°, corresponding to a minimum distance from west Newfoundland to the southern margin of Laurentia of ca. 1500 km. By this time, the Laurentian margin was positioned at about 15°S (Van der Voo *et al.* 1991). Trench & Torsvik (1992) estimated that the latitudinal distance between East Avalonia and the peri-Laurentian terranes was reduced from ca. 5000 to 3300 km between the late Tremadocian and latest Darriwilian, and placed the Welsh Basin of East Avalonia at 45°S in the latest Darriwilian – earliest Sandbian, based on palaeomagnetic studies of volcanic rocks from this basin.

The Iapetus Ocean closed due to northeastward drift of Ganderia and Avalonia towards latitudes of about 11–19°S in the Late Ordovician – mid Silurian (van Staal *et al.* 2012), which resulted in the collision of a number of terranes and microcontinents and ultimately the formation of the North American Appalachians and the northwest European, Scandinavian and Greenland Caledonides. The first sign of initial closure of the Iapetus Ocean was recorded in the rocks from the north and south margins of the ocean by the onset of volcanic activity in association with subduction of oceanic crust (Tietzsch-Tyler & Sleeman 1994). In southeast Ireland, evidence of this event is observed in the arc-associated volcanic rocks of the Lower Ordovician Maulin Formation (Ribband Group) underlying the Duncannon Group and separated from this by a hiatus (Tietzsch-Tyler & Sleeman 1994; Harper & Parkes 2000; Key *et al.* 2005). The collisions involving Ganderia and Avalonia commenced with the arc–arc collision of the western part of the Popelogan-Victoria arc and the Red Indian Lake arc, which at this time formed the leading edge of the Laurentian plate, during the Taconic Orogeny 3 (460–450 Ma) (van Staal *et al.* 2009). This was followed by the accretion of the active leading edge of Ganderia to Laurentia and progressive closure of the Tetagouche-Exploits back-arc basin during the Salinic Orogeny (450–423 Ma) and was concluded in the late Silurian–Early Devonian (421–400 Ma) by the Acadian orogeny involving accretion of Avalonia to Laurentia (van Staal *et al.* 2009). The arc–arc collision between the western Popelogan-Victoria arc and Red Indian Lake arc is today represented by the Red Indian Line, which is the principal Iapetan suture in the northern Appalachians (van Staal *et al.* 2009; Pollock *et al.* 2012).

Ireland

The presence of an Ordovician volcanic arc in southeast Ireland and the English Lake District have been the subject of discussion for many years (e.g. Phillips *et al.* 1976; Stillman 1986; Tietzsch-Tyler & Sleeman 1994; Sleeman & McConnell 1995; Key *et al.* 2005; Brenchley *et al.* 2006). The Anglo-Irish part of the Popelogan-Victoria arc is represented by the present-day Longford-Down Inlier in Ireland and England (Fig. 1). Earlier workers referred to the island arc as the Irish Sea Horst (Williams 1969; Bevins *et al.* 1992; Brenchley *et al.* 2006), the Irish Sea Landmass (Fitton & Hughes 1970; Phillips *et al.* 1976) or the Longford-Down arc (van Staal *et al.* 1998).

Ireland was formed by accretion of at least seven arc terranes within the Iapetus Ocean during the Ordovician–Silurian orogenies (Harper & Parkes 2000; see Fig. 1). Faunal evidence from Harper & Parkes (1989), Parkes (1992) and Harper *et al.* (2013) suggests that the South Mayo Terrane and the Irish–Scottish Midland Valley and Southern Uplands terranes occupied a peri-Laurentian

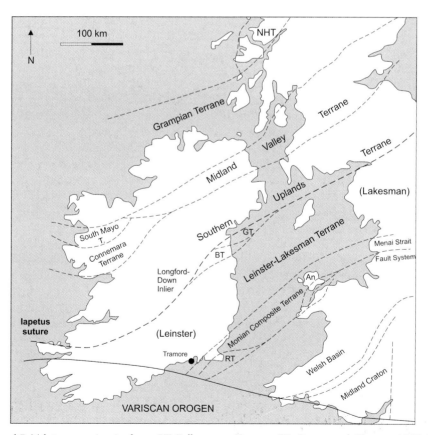

Fig. 1. Map of Irish and British terranes. An, Anglesey; BT, Bellewstown Terrane; GT, Grangegeeth Terrane; NHT, Northern Highlands Terrane; RT, Rosslare Terrane. Modified from Woodcock (2000) and Waldron *et al.* (2014).

position during the Ordovician and were associated with the Red Indian Lake volcanic arc by the northern border of the Iapetus Ocean. A number of workers applying non-faunal data (e.g. van Staal *et al.* 1996, 1998, 2009, 2012; Woodcock 2000; Collins & Buchan 2004; Valverde-Vaquero *et al.* 2006; Pollock *et al.* 2012; Pothier *et al.* 2015; Waldron *et al.* 2014) associated the Leinster–Lakesman, Monian Composite (Anglesey, Rosslare), Bellewstown and Grangegeeth terranes with the leading edge of Ganderia (PVA-TEB) by the southern border of the Iapetus Ocean. Together, these terranes make up present-day Ireland and the Iapetus suture marks the boundary between the rocks formed in the north and south Iapetus Ocean (Fig. 1).

The Leinster Terrane

The Irish part of the Leinster–Lakesman Terrane is termed the Leinster Terrane. It makes up most of the southern part of Ireland (Fig. 1) and covers an area of 24 000 km^2 (Murphy *et al.* 1991). The Leinster Terrane is bounded by the Irish Southern Uplands Terrane and the Bellewstown Terrane to the north and northeast, respectively, along major fault lines running from Dingle in the west to Bellewstown in the east and by the Monian Composite Terrane represented by the Rosslare Terrane to the southeast (Murphy *et al.* 1991; Harper & Parkes 2000; Woodcock 2000; Waldron *et al.* 2014) (Fig. 1). The English part of the Leinster–Lakesman Terrane is termed the Lakesman Terrane. It extends northeast from Ireland through England along the southern border of the Scottish Southern Uplands Terrane (Bluck *et al.* 1992; Woodcock 2000). Max *et al.* (1990) divided the Leinster Terrane into five separate terranes. Harper & Parkes (2000) noted, however, that the stratigraphical development across the terrane is broadly similar and differences in facies may be related to lateral facies changes rather than significant geographical separation.

Docking history

The amalgamation of the Irish terranes may have commenced between the early Cambrian and the Floian with the docking of the Monian Composite Terrane, including the Anglesey and Rosslare terranes, onto the Leinster–Lakesman part of Ganderia (van Staal *et al.* 1998; Valverde-Vaquero *et al.* 2006; Pothier *et al.* 2015). Pothier *et al.* (2015) suggested that Monian deformation juxtaposed a portion of Ganderia, probably represented by present-day Anglesey, against the margin of the Welsh Basin in East Avalonia along the present-day Menai Strait

Fault System (Fig. 1) causing an influx of 'Monian' detritus into the Welsh Basin by the Tremadocian. Following van Staal *et al.* (1998, 2012), the Welsh Basin gradually became separated from the Leinster–Lakesman and Monian Composite terranes by the initiation and continued development of the Tetagouche-Exploits backarc basin from the Floian and onwards. Derivation of sediment from the Monian Composite Terrane into the Welsh Basin continued at least until the Hirnantian (Pothier *et al.* 2015). The Irish Southern Uplands and Bellewstown terranes amalgamated with the northern and northeastern margin of the Leinster Terrane, respectively, in the Llandovery to Wenlock during the Salinic Orogeny (450–423 Ma), in turn followed by amalgamation of the Irish Midland Valley and Grampian terranes (Murphy *et al.* 1991).

The Tramore area

The Tramore Limestone Formation was deposited in a shallow to deep shelf environment on the southeastern Leinster Terrane which deepened westwards into the local Leinster Basin (Fig. 2), into which the Dunabrattin Shale and Limestone formations were deposited (Phillips *et al.* 1976; Key *et al.* 2005). Most of the sedimentary successions belonging to the Duncannon Group represented a period of local relative quiescence before the extensive volcanism, which dominated the Caradoc volcanogenic Carrighalia and Campile formations (Brück *et al.* 1979; Carlisle 1979; Key *et al.* 2005). Most of the upper Darriwilian – lower Sandbian Tramore Limestone Formation was not directly affected by volcanic activity except for the lower and upper parts of the formation. The upper Darriwilian Dunabrattin Shale Formation and the lower part of the lower Sandbian Dunabrattin Limestone Formation accumulated in a volcanically active area of the Leinster Basin. Volcanic activity in this area subsided during deposition of the upper part of the Dunabrattin Limestone, but commenced again and became widespread in southeast Ireland during deposition of the middle Sandbian – lower Katian Carrighalia and Campile formations (Carlisle 1979; Harper & Parkes 2000). The lower Katian (late Caradoc) Raheen contains an abundant deeper-water fauna, dominated by *Onniella* and *Sericoidea* (Harper *et al.* 2016). The chemical composition of the Leinster volcanics including basaltic, andesitic and rhyolitic rocks suggests that the terrane was likely located in a transitional position between the Popelogan-Victoria arc and Tetagouche-Exploits backarc basin (Parkes 1992; Tietzsch-Tyler & Sleeman 1994).

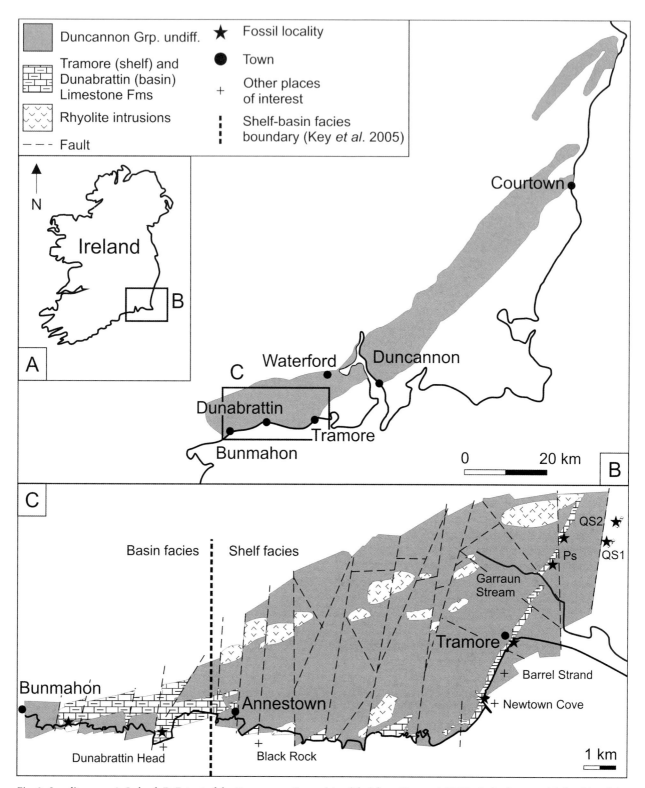

Fig. 2. Locality map. **A**, Ireland. **B**, Extent of the Duncannon Group (simplified from Key *et al.* 2005). **C**, Study area with fossil localities (simplified from Carlisle 1979). Shelf-basin facies boundary from Key *et al.* (2005). Ps, Pickardstown sections; QS1, Quillia Section 1; QS2, Quillia Section 2.

Material

The brachiopod material described in the present study was collected by Carlisle in the 1970s in an area 10 km SW of Waterford city extending from Tramore in the east to Bunmahon in the west. The area is located on the southern margin of a belt of Lower Palaeozoic rocks, which spans southeast Ireland (Fig. 2). The brachiopods were collected from the contemporary shallower-water Tramore and deep-water Dunabrattin Limestone formations.

Carlisle (1979) defined and described the stratigraphical units from which the genera were recorded but most of the samples were not labelled according to locality, formation and level within the stratigraphical column. The present study used Carlisle's personal notes and her 1979 paper to organize the brachiopod occurrences and arrived at three groupings, which define three separate palaeobiogeographical localities in the analyses: Tramore Limestone, shelf-facies brachiopods of late Darriwilian age (Units 1–2); Tramore Limestone, shelf-facies brachiopods of early Sandbian age (Units 3–5 and the barren Unit 6); and Dunabrattin Limestone, outer shelf to basin facies brachiopods of early Sandbian age (Units I–IV).

The collection is currently deposited at the Natural History Museum of Denmark, but will later be transferred to the Natural History Museum, London. During severe flooding in the museum and subsequent relocation of the invertebrate collections, a small number of the figured specimens were mislaid. These specimens are indicated on the plate descriptions and if they do come to light, will be assigned NHM numbers and transferred to the collections in the Natural History Museum, London.

Stratigraphy and fauna

The Tramore and Dunabrattin Limestone formations represent local sedimentary developments in the lower parts of the lower middle Darriwilian to lower Katian Duncannon Group, which comprises a suite of limestones, black shales and basaltic, andesitic and rhyolitic arc volcanics. Both formations are dominated by calcareous clastic sediments interbedded with rhythmically developed, massive to nodular limestone bands. Their stratigraphical relationships to other Duncannon Group formations within and near the study area are shown in Figure 3.

The Duncannon Group occurs in a band trending northeastwards from Dungarvan, County Waterford, towards the south County Wicklow (Key *et al.*

2005) (Fig. 2B). The successions cropping out in the study area (Fig. 2C) are dominated by the southern development of the Duncannon Group comprising a variety of volcanic and volcaniclastic facies (Harper & Parkes 2000). Harper & Parkes (2000) noted that the volcanogenic Bunmahon Formation in the western part of the study area possibly correlates with the Dunabrattin Shale as well as the lower part of the Tramore Limestone Formation, which is supported by this study. The Tramore Limestone Formation and the contemporary Bunmahon and Tramore Shale formations form the oldest parts of the Duncannon Group reaching into the Abereiddian Stage of the Llanvirn Series (Fig. 3). Many local stratigraphies have been described in the Geological Survey of Ireland's 1:100 000 map compilations for the region (e.g. McConnell *et al.* 1994; Tietzsch-Tyler & Sleeman 1994; Sleeman & McConnell 1995).

The age of the various formations belonging to the Duncannon Group is constrained by microfossils, graptolites, brachiopods and trilobites (see Harper & Parkes 2000). These authors also discussed stratigraphy and correlation of the formations.

Age of the Tramore Limestone Formation

Previous studies have assigned the Tramore Limestone Formation to the lower Sandbian based on shelly faunas (Williams 1976; Brenchley *et al.* 1977; Carlisle 1979; Parkes 1994), as well as to the late mid Darriwilian based on conodonts representing the *Eoplacognathus lindstroemi* Subzone of the *Pygodus serra* Zone (Bergström & Orchard 1985). In a revision of Ordovician chronostratigraphy by Fortey *et al.* (1995), the formation was suggested as ranging in age from the latest Darriwilian to the early Sandbian (Llandeilian–Aurelucian). This was also applied by Owen & Parkes (2000) following Bergström & Orchard (1985). Further refinements tabulated in Harper & Parkes (2000) indicated that the formation ranges in age from the upper part of the upper Darriwilian *murchisoni* Zone to the lower Sandbian *gracilis* Zone. Within the Tramore Limestone Formation, the Darriwilian – Sandbian boundary is likely located between lithological Units 2 and 3 or in the lowermost part of Unit 3. This has not been confirmed by graptolite or conodont biostratigraphy but Carlisle (1979) recorded a brachiopod fauna similar to that in Unit 3 of the Tramore Limestone Formation, immediately above the lower boundary of the Dunabrattin Limestone Formation, which was correlated by Harper & Parkes (2000) with the Darriwilian – Sandbian boundary. Hence, this study places the Darriwilian – Sandbian boundary at the

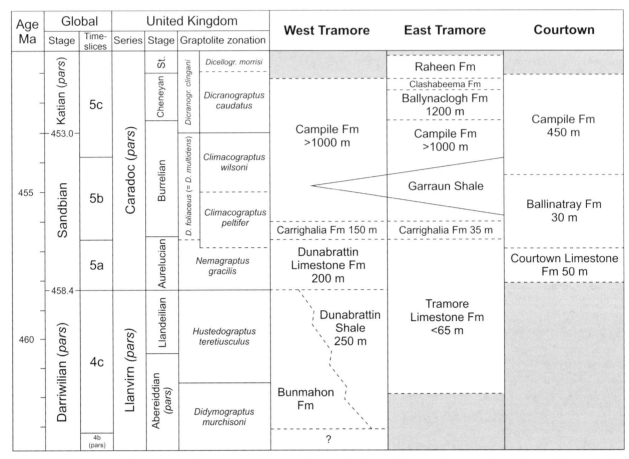

Fig. 3. Correlation chart showing formations of the Duncannon Group including the Tramore Limestone Formation and its correlatives for selected outcrop areas in southeast Ireland. Chronostratigraphy, global stages, British series and stages and graptolite zonation are redrawn from Cooper & Sadler (2012); time-slices are redrawn from Webby *et al.* (2004); and stratigraphical sections, their biostratigraphical correlation and fossil content are redrawn from Harper & Parkes (2000). Abbreviations: *D., Diplograptus; Dicellogr., Dicellograptus; Dicranogr., Dicranograptus;* St., Streffordian.

boundary between lithological Units 2 and 3, within the Tramore Limestone Formation. The Tramore Limestone is overlain by the thin, middle Sandbian Carrighalia Formation which is in turn overlain by the middle Sandbian – lowermost Katian Campile Formation (Stillman 1978; Carlisle 1979; Harper & Parkes 2000).

Tramore and Dunabrattin/Bunmahon sections

The Tramore Limestone Formation crops out between Tramore Bay in the east and Ballydowane Bay in the west, approximately 8.5 km west of Bunmahon town (Key *et al.* 2005) (Fig. 2). The exposures are excellent along the 30- to 60-m-high sea cliffs although they are only accessible by trenched gullies (Wyse Jackson *et al.* 2001; Key *et al.* 2005). The Tramore Limestone grades from a shallower shelf facies in the east to a deeper basinal facies, the Dunabrattin Limestone Formation, in the west with maximum thicknesses of 65 and 450 m, respectively

(Carlisle 1979; Harper & Parkes 2000; Key *et al.* 2005). The transition from shelf to basinal facies occurs between Black Rock and Dunabrattin Head (Key *et al.* 2005) (Fig. 2).

The stratigraphy of the Tramore Limestone Formation and its basinal correlatives was described and measured by Carlisle (1979) from the type sections and other exposures. The Tramore Limestone is primarily described from its type section in the Barrel Strand area, the Dunabrattin Limestone from its type section at Dunabrattin Head and the Dunabrattin Shale Formation from its type section along 1500 m of the foreshore east of Dunabrattin Head and from other sections near Bunmahon town (Fig. 2). The measured sections are shown on the logs in Figure 4. Brachiopods were collected at the fossil localities shown in Figure 2.

Carlisle (1979) divided the Tramore Limestone Formation into informal units and fauna associations and correlated the Tramore Limestone with the Dunabrattin–Bunmahon stratigraphy based on fauna associations. This correlation is mainly

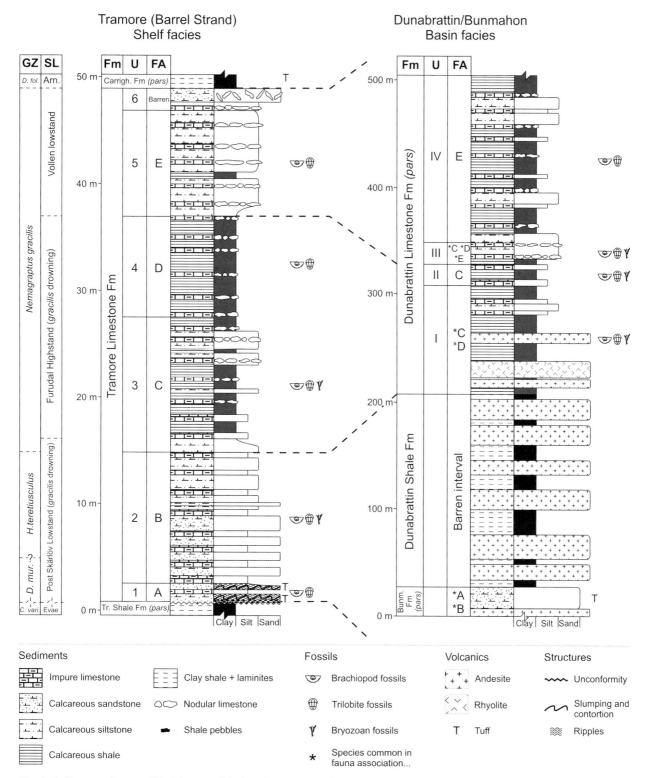

Fig. 4. Sedimentary logs modified from Carlisle (1979). GZ = British graptolite zonation (after Harper & Parkes 2000). SL = Sea-level changes correlated with the likely global events interpreted for Baltica by Nielsen (2004); Evae, Evae Drowning Event; Arn., Arnestad Drowning Event; U, units; FA, fauna association (reinterpreted from Carlisle 1979); Fm, formation. *Species common in the given fauna association of the Tramore Limestone. Graptolites: *C. vari.*, *Corymbograptus varicosus*; *D. mur.*, *Didymograptus murchisoni*; *H. teretiusculus*, *Hustedograptus teretiusculus*; *D. fol.*, *Diplograptus foliaceus*. Formations: Tr. Shale Fm, Tramore Shale Formation; Carrigh. Fm, Carrighalia Formation; Bunm. Fm, Bunmahon Formation.

followed in this study but supplemented by lithological correlation and the graptolite zonation from Harper & Parkes (2000), therefore providing a different correlation of the units than originally suggested by Carlisle (1979).

Tramore Shale Formation – Tramore area

The Tramore Shale Formation is a local correlative of the upper part of the mainly Lower Ordovician Ribband Group and it is separated from the Tramore Limestone Formation by a marked disconformity (Harper & Parkes 2000). Carlisle (1979) recorded exposures of the formation from the east Tramore area: the formation comprises over 1300 m of highly deformed unfossiliferous dark grey shale with subordinate grey/brown siltstone and sandstone. Massively bedded, strongly cleaved, tuffaceous shale containing numerous flattened lenticular lapilli crops out in the region of Perry's Bridge and represent the earliest volcanic activity in the area (Carlisle 1979). The top of the formation is eroded in this area (Carlisle 1979) and Harper & Parkes (2000) described a package of slumped units between the Tramore Shale and Limestone probably masking a disconformity. The depositional disturbance was considered by Sleeman & McConnell (1995) to have been caused by slumping alone and not representing a significant time period. If the slumped units represent a major hiatus, the time interval between the two formations may represent up to 12 Ma and range from the uppermost *Corymbograptus varicosus* Zone to the uppermost *Didymograptus murchisoni* Zone as inferred by Harper & Parkes (2000).

Tramore Limestone Formation – type section at Barrel Strand, Tramore area

The informal units of Carlisle (1979) are used here. The units, referred to as Unit 1–6 in ascending order, are characterized by their lithology and fauna associations (see Fig. 4 and Table 1) and the unit boundaries are coincident with those of the fauna associations.

Lithology
Unit 1, the lowermost unit, is approximately 2 m thick and consists of slumped, calcareous and tuffaceous sandstone with contorted siltstone ribs. Shale pebbles up to 4 cm long are recorded from the lower parts of the unit (Carlisle 1979). The upper boundary of Unit 1 is characterized by the transition from a gradational, tuffaceous calcareous sandstone layer to an impure limestone band followed by calcareous sandstone without tuff as well as a shift from Fauna Association A to B (Table 1).

Unit 2 is approximately 13 m thick and composed of calcareous sandstone in the lower part changing to calcareous siltstone in the upper part. Both sediment types are interrupted by rhythmically occurring impure limestone bands up to 20 cm thick formed by concentrations of shelly fauna and comprising about 35% of the unit thickness (Mitchell *et al.* 1972; Carlisle 1979). The lower boundary is defined at the boundary between the uppermost tuffaceous sandstone of Unit 1 and the lowermost impure limestone band of Unit 2. In the lower part of the unit, some limestone bands make up coquinas of disarticulated *Colaptomena* shells and a sample comprising a coquina of several layers of oriented *Sowerbyella* (*Sowerbyella*) valves in a calcareous siltstone was probably collected from the upper part of this unit (see the systematics section). The fauna is dominated by filter feeders and includes only one common trilobite species *Calyptaulax jamesii*. The fauna has been somewhat exposed to sorting and reorientation of valves but is otherwise fairly well preserved. The upper boundary of Unit 2 is characterized by the transition from an impure limestone band to a gradational calcareous siltstone layer followed by an impure limestone band and calcareous shale as well as a shift from Fauna Association B to C (Table 1).

Unit 3 is approximately 14 m thick and consists of calcareous shale and siltstone. Impure limestone bands up to 15 cm thick comprising about 25% of the unit thickness occur rhythmically (Mitchell *et al.* 1972; Carlisle 1979). The limestone bands make up coquinas formed by shell beds in the lower part of the unit. Upwards these bands become nodular and concomitantly the content of the shelly fauna declines (Carlisle 1979). The shelly fauna is fragmented and sorted. The upper boundary of Unit 3 is characterized by the gradational transition from calcareous shale to an impure nodular limestone band a little more than 1 m above a 3-m-thick horizon of dominantly calcareous siltstone as well as a shift from Fauna Association C to D (Table 1).

Unit 4 is approximately 10 m thick and constitutes a dark shale with rhythmically occurring impure nodular limestone bands up to 10 cm thick, which comprise up to 12% of the unit thickness (Mitchell *et al.* 1972; Carlisle 1979). A sparse fauna dominated by small trilobites is concentrated in the limestone bands and the preservation of the shelly fauna is excellent (Carlisle 1979). The upper boundary of Unit 4 is characterized by the transition from dominantly calcareous shale to dominantly calcareous siltstone. The boundary is defined between an

Table 1. Lithological units and fauna associations.

	Unit 1 FA-A	Unit 2 FA-B	Unit 3 FA-C	Unit 4 FA-D	Unit 5 FA-E
Brachiopods					
Acanthocrania			c		
Atelelasma			c		
Colaptomena		vc			
Dactylogonia		c	c		
Glyptorthis	o	o	c		
Grorudia				c	c
Hesperorthis	o	c	o		
Howellites			c		
Isophragma					c
Leptellina			c		
Leptestiina			o		o
Paurorthis		o	c		
Platystrophia			c		
Hibernobonites n. gen.			c		
Salopia			c		
Sowerbyella (*Sowerbyella*)		vc	o	o	o
Sulevorthis		o			
Trilobites (Carlisle 1979)					
Ampyx				o	o
Botrioides					vc
Calyptaulax brongiarti				c	c
Calyptaulax jamesii	o	c			
Deacybele					o
Eirelithus	o	o		c	
Encrinuroides				c	c
Isotelus				o	o
Oedicybele					o
Platycalymene					o
Pliomerops					o
Remopleurides sp. A				o	
Remopleurides sp. B					o
Tramoria				o	
Trinodus				c	o
Bryozoans (Carlisle 1979)					
Bryozoa undiff.		c	c		
Monticuliporina		c	c		

FA, fauna association; o, occasional; c, common; vc, very common. Abundances are from Carlisle (1979). The terms 'occasional', 'common' and 'very common' are relative.

impure nodular limestone band and a gradational layer of calcareous siltstone with a shift from Fauna Association D to E (Table 1).

Unit 5 is approximately 11 m thick and composed almost entirely of calcareous siltstone with rhythmically occurring impure nodular limestone bands up to 12 cm thick formed by concentrations of trilobite remains (notably *Botrioides*). These bands comprise up to 20% of the unit thickness (Mitchell *et al.* 1972; Carlisle 1979). Calcareous shale is only a minor component of the unit. The shelly fauna is well preserved. The upper boundary of Unit 5 is characterized by the transition from undisturbed calcareous siltstone to tuffaceous sandstone with broken

disoriented limestone bands. The boundary is set between a thick calcareous siltstone and a thin impure nodular limestone band followed by a thin calcareous siltstone layer and the shift from Fauna Association E to the barren sediments of Unit 6 (Table 1).

Unit 6, the uppermost unit of the Tramore Limestone Formation is approximately 2 m thick and consists of tuffaceous sandstone with broken, disoriented limestone layers (Carlisle 1979). This unit seems to be barren. The Tramore Limestone is conformably overlain by the Carrighalia Formation and the boundary is characterized by a lithological change from the tectonically disturbed, tuffaceous calcareous sandstone of Unit 6 to tuffaceous calcareous shale of the lowermost Carrighalia Formation.

Fauna associations

The fauna associations of the Tramore Limestone Formation were defined based on the type section in the Barrel Strand area by Carlisle (1979). They are reproduced here but modified according to the new taxonomical revisions. The trilobite and bryozoan lists in Table 1 are from Carlisle (1979) and have not been revised. The fauna associations were originally defined by the co-occurrence of the genera and species in Table 1. The abundances were not quantified by Carlisle (1979) and, hence, the terms 'occasional', 'common' and 'very common' are relative. The occurrence of all brachiopod species recorded from the units and sections of the Tramore and Dunabrattin limestones are listed in Tables 2 and 3, respectively, but not quantified as the samples in the collection were not labelled according to stratigraphy. The stratigraphical occurrences of the brachiopod species were inferred from Carlisle (1979) and Carlisle's personal notes from the 1970s as mentioned earlier.

Fauna Association A (FA-A) was recorded from Unit 1. The fauna is sparse and dominated by the occasional presence of the brachiopods *Glyptorthis crispa* and *Hesperorthis leinsterensis* n. sp., and the trilobites *C. jamesii* and *Eirelithus*.

Fauna Association B (FA-B) was recorded from Unit 2. The overall abundance and diversity of the fauna is higher than in FA-A. Common to very common in this association are the brachiopods *Colaptomena auduni* n. sp., *Dactylogonia costellata* n. sp., *H. leinsterensis* n. sp. and *Sowerbyella* (*Sowerbyella*) *antiqua* as well as the trilobite *C. jamesii* and bryozoans.

Fauna Association C (FA-C) was recorded from Unit 3. The fauna is very abundant and diverse and dominated by benthic filter feeders. The common fossils of this association are brachiopods *Acanthocrania* sp., *Atelelasma longisulcum* n. sp., *D. costellata* n. sp., *G. crispa*, *Howellites hibernicus* n. sp., *Leptellina llandeiloensis*, *Paurorthis* aff. *Paurorthis parva*, *Platystrophia tramorensis* n. sp. (*P.* aff. *Platystrophia sublimis* in the Dunabrattin Limestone), *Hibernobonites* n. gen. *filosus* and *Salopia gracilis* as well as bryozoans.

Fauna Association D (FA-D) was recorded from Unit 4. The fauna is sparse and dominated by trilobites, notably *Calyptaulax brongiarti*, *Eirelithus*, *Encrinuroides* and *Trinodus*. Brachiopods are scarce and occur only occasionally but Carlisle noted that *Grorudia grorudi* is common and associated with the

siltstones of the unit. This study has added the common occurrence of *G. grorudi* to the definition of FA-D (Table 1).

Fauna Association E (FA-E) was recorded from Unit 5. The fauna is dominated by an abundant and diverse trilobite fauna with common to very common *Botrioides*, *C. brongiarti* and *Encrinuroides*. Commonly occurring brachiopods include the two species *G. grorudi* and *Isophragma parallelum* n. sp. Otherwise brachiopods are rare and the diversity is low.

Tramore Limestone Formation – other sections

Northeast of Tramore town, several inland exposures of calcareous sandstone and grit belonging to the Tramore Limestone Formation occur.

Pickardstown sections (Ps)

At Pickardstown several small outcrops and loose blocks of calcareous sandstone are packed with a fauna typical of FA-B, notably the brachiopods *Colaptomena*, *Dactylogonia*, *Hesperorthis*, *Sowerbyella* (*Sowerbyella*) and the trilobite *C. jamesii* (Carlisle 1979; see Table 2). The section may be correlated with Unit 2 of the type section and Quillia Section 1.

Quillia Section 1 (QS1)

At Quillia, 3 m of sandy limestone packed with large specimens of the brachiopod *Colaptomena* and less common *Sowerbyella* (*Sowerbyella*) as well as the trilobite *C. jamesii* is suggestive of a shell bank (Carlisle 1979) (see Table 2). Based on the fauna, the QS1 may be correlated with Unit 2 of the type section and to the Pickardstown sections.

Quillia Section 2 (QS2)

Two exposures of calcareous sandstone and grit containing a sparse fauna are observed to the northeast of Quillia. The fauna includes genera common in Units 2 and 3 of the type section (Tables 1 and 2), notably the brachiopods *Dactylogonia*, *Glyptorthis*, *Hesperorthis*, *Paurorthis* and *Sowerbyella* (*Sowerbyella*), but it also includes *Grorudia*, more commonly associated with the siltstones of Unit 4 and two other genera *Glyptambonites* and *Productorthis* unknown from the other Tramore sections (Carlisle 1979). Carlisle (1979) interpreted this as being a shallower-water equivalent of the upper parts of the type section. The first appearance of *Grorudia* in the type section is recorded, however, in Unit 3 in the middle part of the section. As all the other brachiopod genera of QS2 were also recorded from Unit 3, the current study correlates QS2 with Unit 3 of the type section.

Table 2. Species occurrence and palaeobathymetry in the Tramore Limestone sections.

	Unit 1 DR 3	Unit 2 DR 3	Unit 3 DR 3–4	Unit 4 DR 4	Unit 5 DR 3	Ps DR 3	QS1 DR 3	QS2 DR 3
Atelelasma longisulcum n. sp.			X					
Colaptomena auduni n. sp.		X				X	X	
Colaptomena pseudopecten?		X						
Dactylogonia costellata n. sp.		X	X					X
Glyptambonites sp.								X
Glyptorthis crispa	X	X	X					X
Grorudia grorudi			X	X	X			X
Hesperorthis leinsterensis n. sp.	X	X	X		X	X		X
Howellites hibernicus n. sp.	X	X	X		X			
Isophragma parallelum n. sp.					X			
Leptellina llandeiloensis		X	X					
Leptestiina derfelensis			X		X			
Paurorthis aff. *parva*	X	X	X					X
Platystrophia tramorensis n. sp.		X	X					
Hibernobonites n. gen. *filosus*		X	X					
Paurorthis aff. *P. parva*								X
Salopia gracilis		X	X		X			
Sowerbyella (S.) antiqua		X	X	X	X	X	X	X
Sulevorthis aff. *S. blountensis*	X	X	X	X				
Tetraphalerella? sp.	?	?						
Valcourea confinis		X	X					

DR, Depth Range zones; Ps, Pickardstown sections; QS1, Section at Quillia; QS2, Section northeast of Quillia.

Bunmahon Formation – Bunmahon/Dunabrattin area

The Bunmahon Formation is volcanogenic and probably correlates with the Dunabrattin Shale and Tramore Limestone formations (Harper & Parkes 2000). Carlisle (1979) recorded the uppermost part of the 'Bunmahon Group' from the area around Bunmahon and Dunabrattin Head. It consists of andesite overlain by approximately 22 m of tuffaceous calcareous sandstone and most likely corresponds to the Bunmahon Formation of Harper & Parkes (2000). It is conformably overlain by the Dunabrattin Shale Formation. The boundary between the Bunmahon Formation and the overlying Dunabrattin Shale Formation is a transition from a tuffaceous calcareous sandstone to laminated clay shale. The uppermost Bunmahon Formation contains taxa occurring in FA-A and commonly occurring in FA-B of the Tramore Limestone Formation (Table 1).

Dunabrattin Shale Formation – Bunmahon/Dunabrattin area

The Dunabrattin Shale Formation comprises a suite of intensely deformed shale and laminites with subordinate volcanics (Carlisle 1979). The type section is a faulted and folded succession up to 250 m thick exposed along 1500 m of the foreshore east of Dunabrattin Head (Carlisle 1979). Here the lower parts of the formation are dominated by laminites revealing load casts; ripple lamination and large and small scale slumps, which may indicate a turbidite origin. The sediments are interspersed by massive tuff units and vesicular andesitic lavas with mixed and disturbed upper and lower junctions. Large-scale slumping and shale breccias in this sequence are evidence of syndepositional instability caused by tectonic activity related to the arc volcanism. The lack of fossils in the Dunabrattin Shale may be related to an unsuitable environment for brachiopods and/or due to diagenesis and metamorphosis. The barren clay shale of the Dunabrattin Shale Formation is overlain by impure limestone bands and calcareous shale assigned to the Dunabrattin Limestone Formation. These calcareous shales contain the commonly occurring taxa of FA-C and FA-D of the Tramore Limestone Formation.

Dunabrattin Limestone Formation – Bunmahon/Dunabrattin area

Carlisle (1979) did not subdivide the formation but described it according to faunal content and lithology and compared it with the fauna associations of the Tramore Limestone Formation. This study subdivides the Dunabrattin Limestone Formation into informal units characterized by their lithology and fauna associations to ease correlations and descriptions. The units are named Unit I to IV and are not lithologically and faunally similar to Units 1–4 of the Tramore Limestone Formation.

Lithology and fauna

The lowermost unit, Unit I, of the Dunabrattin Limestone Formation is approximately 42 m thick and composed of calcareous shale and siltstone with bands of impure limestone enriched in trilobite remains (Carlisle 1979). At the quay on the northern side of Dunabrattin Head, the base of the formation is characterized by the appearance of the first limestone bands (Carlisle 1979). The brachiopod fauna contains taxa commonly occurring in FA-C and FA-D of the Tramore Limestone Formation. The upper boundary of Unit I is characterized by a shift in the fauna from containing the commonly occurring taxa of FA-C and FA-D to comprising the fauna of FA-C (Tables 1 and 3). The overlying Unit II is lithologically similar to the uppermost part of Unit I although it does not contain calcareous silt layers. The boundary is placed between a calcareous shale layer and an overlying impure limestone band.

Unit II is approximately 20 m thick and consists of calcareous shale with impure limestone bands becoming more frequent than in Unit I and contains a shelly fauna dominated by benthic filter feeders (Carlisle 1979). The brachiopod fauna is similar to FA-C of the Tramore Limestone Formation. The upper boundary of the unit is characterized by a shift in the fauna from FA-C to the first appearance of the taxa commonly occurring in FA-E (Tables 1 and 3). The boundary is placed between the last, rather thick, massive impure limestone band and a calcareous shale followed by a nodular impure limestone band.

Unit III is approximately 20 m thick and is dominated by calcareous shale in the lowermost part and calcareous siltstone in the main upper part of the unit with a few impure nodular limestone bands in the siltstone. The shelly fauna is dominated by taxa commonly occurring in FA-C, FA-D and FA-E of the Tramore Limestone Formation (Tables 1 and 3). The upper boundary of the unit is placed between an impure nodular limestone band and a layer of siltstone where the brachiopod fauna shifts from being dominated by the commonly occurring taxa of FA-C, FA-D and FA-E to being similar to FA-E of the Tramore Limestone (Tables 1 and 3).

The uppermost unit of the exposed section, Unit IV, comprises up to 150 m of calcareous shale with less prominent impure nodular limestone bands and a few siltstone layers, 4–8 m thick. The limestone bands appear to occur rhythmically as in the Tramore Limestone Formation but this was not described by Mitchell *et al.* (1972) who focused on the Tramore and Courtown Limestone formations. The upper part of the unit contains a thick succession of cleaved, calcareous siltstone with discontinuous impure limestone layers formed by trilobite coquinas. The shelly fauna is similar to FA-E of the Tramore Limestone Formation (Tables 1 and 3). The top of the formation was not exposed.

Correlation

The exposed upper part of the Bunmahon Formation is correlated with Unit 1 and 2 of the Tramore Limestone Formation from the type section based on the presence of taxa commonly occurring in FA-A and FA-B (Table 1). Carlisle (1979) correlated the Bunmahon Formation with the hiatus between the Tramore Limestone and Shale formations but did not discuss the basis for this interpretation.

The Dunabrattin Shale Formation is barren of fossils and, hence, difficult to correlate. It lies above the

Table 3. Species occurrence and palaeobathymetry in the Dunabrattin Limestone sections.

	Unit I DR 4–6	Unit II DR 4–6	Unit III DR 4–5	Unit IV DR 4–5
Atelelasma longisulcum n. sp.	x	x	x	
Dactylogonia costellata n. sp.	x	x	x	
Glyptorthis crispa	x	x	x	
Grorudia grorudi			x	x
Hesperorthis leinsterensis n. sp.		x		
Howellites hibernicus n. sp.	x	x	x	x
Isophragma parallelum n. sp.			x	x
Leptellina llandeiloensis	x	x	x	
Leptestiina derfelensis		x		x
Paurorthis aff. *P. parva*	x	x	x	
Platystrophia aff. *P. sublimis*	x	x	x	
Hibernobonites n. gen. *filosus*	x	x	x	
Salopia gracilis	x	x	x	
Sowerbyella (S.) *antiqua*		x		x

DR, Depth Range zones.

Bunmahon Formation, containing taxa commonly occurring in Unit 1 and 2 of the Tramore Limestone (FA-A + B; Table 1), and below the Dunabrattin Limestone containing taxa commonly occurring in Unit 3 and 4 of the Tramore Limestone (FA-C + D; Table 1). The boundary between the Dunabrattin Shale and Dunabrattin Limestone is here tentatively correlated with the boundary between Unit 2 and 3 of the Tramore Limestone based on the taxa commonly co-occurring in FA-C and the first appearance of *Atelelasma*. This boundary corresponds to the Darriwilian – Sandbian boundary as inferred from Harper & Parkes' (2000) biostratigraphical correlation of the boundary between the Dunabrattin Shale and Limestone formations.

The lower boundary of the Dunabrattin Limestone Unit III is correlated with the lower boundary of the Tramore Limestone Unit 5 based on the taxa commonly co-occurring in FA-E (Table 1) and the first appearance of *Isophragma*. Dunabrattin Limestone Units I and II are therefore contemporary with Tramore Limestone Units 3 and 4. Carlisle (1979) correlated the lower boundary of the Dunabrattin Limestone Unit III with the lower boundary of Tramore Limestone Unit 4 but did not give an explanation.

The top of the Dunabrattin Limestone Formation was not exposed, and hence, it was not possible to correlate the boundary between the Tramore Limestone Formation and the Carrighalia Formation with the Dunabrattin/Bunmahon section. Carlisle (1979) correlated the uppermost part of the exposed Dunabrattin Limestone with the lower boundary of the unfossiliferous Unit 6 of the Tramore Limestone without explanation.

Depositional setting and ecology

Deposition of the Duncannon Group commenced with a global rise in sea level, the *gracilis* drowning, culminating in a maximum highstand recorded as the Furudal Highstand from the Baltic area by Nielsen (2004). This resulted in a deepening of the depositional basin and transgression to the east over the eroded basin margins of the Tramore Shale Formation (Carlisle 1979) possibly accompanied by local subsidence. The upper parts of the Tramore and Dunabrattin Limestone were deposited during a period of global regression correlated with the Vollen Lowstand *sensu* Nielsen (2004). The rhythmic limestone bands of the Tramore and Dunabrattin Limestones indicate cyclic deposition with recurring periods of reduced clastic input. This has never been properly assessed and the sections should be measured in detail to analyse what type of cycles they represent.

Local events

The varying lithologies and sedimentary structures of the studied sections as well as the changing faunal content and preservation of shells both throughout the study area and within specific successions show that the depositional setting and ecology fluctuated laterally and vertically. The Tramore and Dunabrattin limestones represent continuous deposition of carbonate-rich sediments throughout the area grading from a shallow shelf facies with a depth range from the intertidal zone to outer shelf in the east (Tramore Limestone, Depth Range zones 1–4; Fig. 5) to outer shelf–slope/basinal facies within Depth Range zones 4–6 in the west (Dunabrattin Limestone).

The Depth Range zones (DRs; Fig. 5) were defined based on lithology, sedimentary structures and preservation of fossils relating to syndepositional mechanical processes in shelf environments and were applied to the localities in the database as a means of comparison of palaeodepth. They are also applicable to carbonate platforms although the different sediment types and structures in these environments have to be taken into account. Some authors have used the Benthic Assemblage zones (BAs) of Boucot (1975) to indicate the palaeodepth and ocean floor environment of their brachiopod faunas. The BAs were defined based on Silurian–Devonian brachiopod assemblages and have to be correlated and modified when applied to other time periods. Applying the physical parameters described above is more straightforward as their appearance is directly related to depositional environment. This relationship has not changed through time as the generic compositions and ecologic preferences of the brachiopod faunas have. The Depth Range zones of this study are comparable to the depth ranges of the Late Ordovician Benthic Assemblage zones of Rasmussen *et al.* (2012) who correlated and modified the BAs of Boucot (1975).

Tramore Limestone Formation

The Tramore Limestone succession of the Barrel Strand type area displays a wide range of depositional environments, some of which can be related to the global sea-level events described from the Baltic area by Nielsen (2004). Local fluctuations in relative sea level could have been caused by filling and evacuation of an underlying magma chamber related to the arc–backarc volcanism resulting in local shallowing due to uplift and local deepening due to subsidence, respectively.

Unit 1 was deposited in a shallow-water, high-energy environment inferred from the silty and sandy sediments and the presence of siltstone ripples. The shale pebbles in the lower parts of the unit suggest that the seafloor of the Barrel Strand area might have been exposed to subaerial erosion before deposition of the Tramore Limestone commenced. The water depth increased and the lower unit was deposited in water depths probably within the upper subtidal zone, above fair weather wave base (shallow DR 3) (Fig. 5). The unit contains tuff indicating volcanic activity in the vicinity of the depositional site, and contortion and slumping of the sediments indicate instability of the seafloor likely related to tectonic activity and/or the deepening. The shelly fauna is of low diversity, probably resulting from the unstable environment and high amounts of sediment suspended in the water column, possibly due to high sedimentation rate, making it difficult for benthic filter feeders to thrive.

The deepening observed in Unit 1 continued through Unit 2, which was deposited under generally shallow-water, moderate-energy conditions, deepening towards the upper part of the unit from shallow subtidal (shallow DR 3) to depths between fair weather wave base and normal storm wave base (deep DR 3) (Fig. 5). This is indicated by a shift from dominantly calcareous sand to calcareous silt about halfway through the unit with at least one

shell bed interpreted as a winnowed storm lag consisting of oriented *Sowerbyella* (*Sowerbyella*) valves (see the systematics section). Sea level and/or clastic input fluctuated during deposition of the unit as indicated by the alternation between thick calcareous sand-/siltstone layers and thinner impure limestone bands. The fauna is dominated by filter feeding benthos (i.e. brachiopods and bryozoans) and includes only one common trilobite species with well-developed compound eyes. The brachiopod fauna is more abundant and diverse than in Unit 1 and concentrated in the limestone bands, indicating that the limestone bands were deposited under generally clear-water conditions with low clastic input. The calcareous sandstones and grit from Pickardstown (Ps) and Quillia (QS1) are shallow-water equivalents of Unit 2, located to the northeast of the Barrel Strand area and were likely deposited in the shallow subtidal zone (shallow DR 3).

The sediments of Unit 3 were deposited during the still ongoing deepening with minor fluctuations in the overall sea level. Depth probably fluctuated between fair weather wave base and mid shelf below normal storm wave base (deep DR 3–shallow DR 4) (Fig. 5). This is inferred by the change from the dominantly sandy and silty sediments of Unit 2 to the dominantly clayey sediments of Unit 3 alternating with impure limestone bands and layers of calcareous silt. The alternating lithologies were caused by fluctuating sea level and/or clastic input and probably also storm related activity.

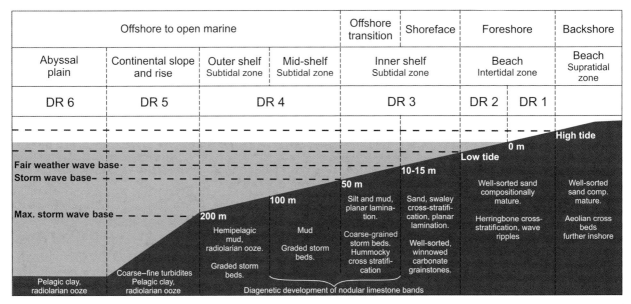

Fig. 5. Standardized coast-shelf-basin profile showing bathymetry, sedimentology, depositional environment in which formation of nodular limestone bands during diagenesis is possible, and Depth Range zones (DRs) in relation to sea level. Sedimentology after Tucker (1991), formation of nodular limestone bands from Möller & Kvingan (1988), bathymetry after Brett *et al.* (1993). The DRs are comparable to the depth ranges of the late Ordovician Benthic Assemblage zones (BAs) of Rasmussen *et al.* (2012).

Fragmentation of small shells and sorting implies the fauna suffered some post-mortem transport, which support the presence of storm related deposition. The diverse and abundant shelly fauna is dominated by benthonic filter feeders concentrated in the limestone bands (Carlisle 1979). The brachiopod fauna reached a maximum level of diversity in the Barrel Strand succession suggesting stable conditions with clear water and sufficient supply of nutrients. The shelly fauna declines upwards in the unit as the limestone bands become nodular (Carlisle 1979) possibly due to diagenesis and pressure dissolution (see Möller & Kvingan 1988). Another factor could be the increasing depth diminishing circulation at the seafloor, which is supported by the presence of a well-preserved, abundant and diverse trilobite fauna and rare filter feeding benthos in the nodular limestone bands of Unit 5. Together with Unit 4, Unit 3 constitutes a period of sea-level highstand, which can be correlated with the *gracilis* drowning (Harper & Parkes 1989; Nielsen 2004).

The exposures of calcareous sandstones and grit northeast of Quillia (QS2) were correlated with Unit 3 based on the brachiopods. The lithology of QS2 indicates, however, that it was deposited under similar shallow-water conditions to Unit 2 and the other northeastern Tramore Limestone sections (Ps and QS1) implying that the shallow-water conditions within shallow DR 3 continued to exist northeast of the type area. Unfortunately, younger exposures of Tramore Limestone northeast of Barrel Strand have not been described, making it impossible to infer if this area remained a shallow-water site while the type area underwent deepening.

The period of maximum depth of deposition during the *gracilis* drowning event is recorded in Unit 4 and corresponds approximately to the outer shelf environment (deep DR 4) (Fig. 5). The unit is exclusively composed of alternating thick layers of calcareous shale and thinner layers of nodular, impure limestone bands, and it contains a sparse fauna concentrated in the limestone bands, which is dominated by trilobites with rare occurrences of brachiopods. The scarcity of filter feeding benthos and excellent preservation in combination with the very fine sediment indicate a relatively deep-water, low-energy, muddy environment possibly with restricted circulation.

At the lower boundary of Unit 5, the sediment quickly grades from calcareous shale to calcareous silt indicating a pronounced shallowing resulting in a change from an outer shelf environment below normal storm wave base (deep DR 4) to an inner

shelf environment between fair weather wave base and normal storm wave base (deep DR 3) (Fig. 5). This sea-level fall may be correlated with the Vollen Lowstand of the Baltic area described by Nielsen (2004). The fauna is dominated by well-preserved trilobites, which are both diverse and abundant and concentrated in the nodular limestone bands. The good preservation of the fauna, the fine sediment and scarcity of filter feeding benthos are suggestive of a low-energy environment with large amounts of sediment suspended in the water column, which could be due to high sedimentary supply.

The tuffaceous sandstone with broken disoriented limestone bands of Unit 6 indicates an unstable, shallow-water environment, probably above fair weather wave base within the subtidal zone (shallow DR 3) (Fig. 5) with renewed volcanic and tectonic activity. Disruption of the limestone bands must have happened after consolidation or cementation of the limestone but before deposition of the undisturbed, laminated clay shales of the lowermost part of the Carrighalia Formation in the Barrel Strand section. This implies the presence of a hiatus between the top of the Tramore Limestone Formation and the lower boundary of the Carrighalia Formation in this area. The change in lithologies indicates that a major deepening must have taken place between the two formations as the clay shale of the lowermost Carrighalia Formation must have been deposited at water depths below normal storm wave base (DR 4–6) (Fig. 5). This was not described by previous studies and the profile should be revisited for further investigation.

Dunabrattin Limestone Formation

The overall sparse fauna and predominance of thick layers of calcareous shale alternating with thin impure limestone bands and a few layers of calcareous silt indicate a deep-water depositional environment with restricted circulation on the outer shelf (DR 4) or possibly even as deep as the slope-basin environment (DR 5–6) through most of the succession (Fig. 5). Carlisle (1979) did not describe sedimentary structures within this formation and the section should be revisited to investigate whether the silt layers show structures related to storm activity or gravity flows (e.g. turbidites). Units III–IV of the Dunabrattin Limestone were correlated with Unit 5 of the Tramore Limestone, which was clearly deposited during a lowstand probably correlating with the Vollen Lowstand. The silt layers of Dunabrattin Units III–IV are thinner (4–8 m) than in Unit 5 of the Tramore Limestone (4–17 m) which, apart from the rhythmically occurring nodular limestone bands

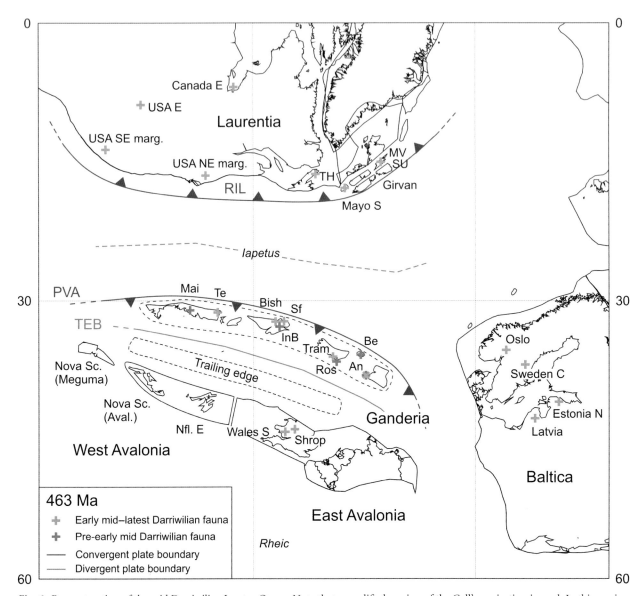

Fig. 6. Reconstruction of the mid Darriwilian Iapetus Ocean. Note that a modified version of the Gall's projection is used. In this version the distance between 0–30 and 30–60°S is equal. An, Anglesey (Monian Composite Terrane); Aval., Avalonia; Be, Bellewstown Terrane; Bish, Bishop's Falls (Central Newfoundland Terrane); Canada E, east Canada; Estonia N, north Estonia; InB, Indian Bay (Central Newfoundland Terrane); Mayo S, South Mayo Terrane; Mai, Maine (Miramichi Terrane); MV, Midland Valley Terrane; Nfl. E, east Newfoundland; Nova Sc., Nova Scotia; PVA, Popelogan-Victoria arc; RIL, Red Indian Lake arc; Ros, Rosslare (Monian Composite Terrane); Sf, Summerford (Central Newfoundland Terrane); Shrop, Shropshire; SU, Southern Uplands Terrane; Sweden C, central Sweden; Te, Tetagouche (Miramichi Terrane); TEB, Tetagouche-Exploits backarc basin; TH, Table Head (west Newfoundland, foreland basin); Tram, Tramore (Leinster Terrane); USA E, east USA; USA NE marg., northeast margin of USA; USA SE marg., southeast margin of USA; Wales S, South Wales.

and one thin calcareous shale layer, consists entirely of calcareous silt. The silt layers of Dunabrattin Units III–IV may represent periods of maximum sea-level lowstand in the study area with 'normal' transport of silt in suspension over the shelf and deposition on the deep inner shelf (deep DR 3). The sediments of Units III–IV are dominated by calcareous shale however, indicating that deposition mainly took place within the outer shelf-slope-basin environment (DR 4–6) and that at least some of the silt layers were likely deposited by storms and gravity

flows on the outer shelf or slope (deep DR 4/DR 5) rather than representing inner shelf conditions.

Palaeobiogeographical reconstructions

A reliable palaeogeographical template on which faunal distributions can be plotted is of crucial importance to any assessment of biogeography. New reconstructions of the plate configuration in

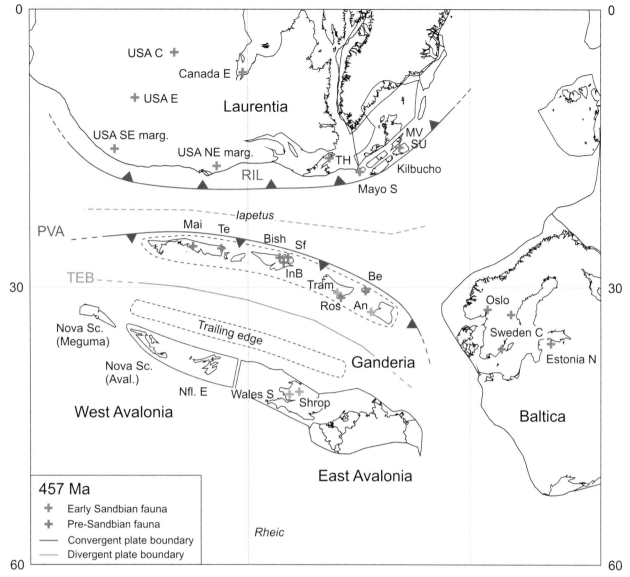

Fig. 7. Reconstruction of the early Sandbian Iapetus Ocean. Note that a modified version of the Gall's projection is used. In this version, the distance between 0–30 and 30–60°S is equal. An, Anglesey (Monian Composite Terrane); Aval., Avalonia; Be, Bellewstown Terrane; Bish, Bishop's Falls (Central Newfoundland Terrane); Canada E, east Canada; Estonia N, north Estonia; InB, Indian Bay (Central Newfoundland Terrane); Mayo S, South Mayo Terrane; Mai, Maine (Miramichi Terrane); MV, Midland Valley Terrane; Nfl. E, east Newfoundland; Nova Sc., Nova Scotia; PVA, Popelogan-Victoria arc; RIL, Red Indian Lake arc; Ros, Rosslare (Monian Composite Terrane); Sf, Summerford (Central Newfoundland Terrane); Shrop, Shropshire; SU, Southern Uplands Terrane; Sweden C, central Sweden; Te, Tetagouche (Miramichi Terrane); TEB, Tetagouche-Exploits backarc basin; TH, Table Head (west Newfoundland, foreland basin); Tram, Tramore (Leinster Terrane); USA C, central USA; USA E, east USA; USA NE marg., northeast margin of USA; USA SE marg., southeast margin of USA; Wales S, South Wales.

the Iapetus Ocean (Figs 6 and 7) and the suggested brachiopod provinces (Figs 10 and 14) and current configurations (Fig. 11) for the Iapetus and Rheic oceans and Tornquist Sea are plotted on palaeogeographical templates based on palaeomagnetic data compilations for Laurentia (Torsvik *et al.* 1990, 1996), Baltica (Torsvik *et al.* 1992), East Avalonia (Trench & Torsvik 1992), Gondwana (Bachtadse & Briden 1990) and the Central Newfoundland Terrane of Ganderia (Van der Voo *et al.*

1991) as well as the new global palaeogeographical reconstructions by Torsvik & Cocks (2013). The palaeomagnetic data do not provide longitudinal constraints and separations between continents are described in terms of latitude only. Hence, longitudinal separation between the studied sites was based on the present faunal data. Ganderian terrane positions and the position of East Avalonia are modified by this study as described below by calculating latitudinal positions from the drift rates and spreading rates given by van Staal *et al.* (2012) for the leading and trailing edges of Ganderia and the Tetagouche-Exploits backarc basin, respectively. The positions of the peri-Laurentian South Mayo, Midland Valley and Southern Uplands terranes are based on biogeographical results from this study and those of Harper & Parkes (1989), Parkes (1992) and Candela & Harper (2010) as well as zircon geochronology by Waldron *et al.* (2014). The placement and orientation of the peri-Gondwanan microplates follows Harper *et al.* (2013). Palaeogeographical positions of the Argentinian sites of Famatina and Cordillera Oriental are from Benedetto (2003a). These are not included in the present reconstructions as they were positioned outside of the main study area. Cordillera Oriental was part of the Northwest Argentina Basin on the (palaeo)mid-northern South American plate (i.e. the present mid-west), whereas Famatina was a peri-South American terrane located along the (palaeo)northern margin of South America (i.e. the present-day western margin).

New reconstructions for the Iapetus Ocean

The new mid Darriwilian and early Sandbian reconstructions of the south Iapetus PVA-TEB terranes, Ganderia and Avalonia (Figs 6 and 7, respectively) are plotted on modified BugPlates maps (Torsvik 2014) using the Gall's projection. The distance on the plots between 30 and 60°S is shortened to correspond with the distance between 0 and 30°S for illustrative purposes. Plate boundaries are modified from van Staal *et al.* (2012), latitudes are calculated as described below, and longitudinal separation is estimated according to the faunal analyses.

Ganderian terrane positions

The position of the PVA-TEB terranes in the reconstructions are based on faunal evidence and the earlier mentioned analyses of Barr & White (1996), van Staal *et al.* (1996, 1998, 2009), Collins & Buchan (2004), Valverde-Vaquero *et al.* (2006), Hibbard *et al.* (2007), Fyffe *et al.* (2009), Pothier *et al.* (2015) and Waldron *et al.* (2014). Portions of the island of Anglesey and the Leinster–Lakesman Terrane have been correlated with the Ganderian part of Newfoundland and New Brunswick (van Staal *et al.* 1996, 1998; Waldron *et al.* 2014) and Pothier *et al.* (2015) placed the Monian Composite Terrane (including Anglesey and the Rosslare Terrane) as an amalgamated part of the Leinster–Lakesman Terrane from at least 500 Ma in the eastern part of Ganderia, which is followed here (Figs 6 and 7). The Central Newfoundland (Bishop's Falls, Indian

Bay and Summerford localities) and Miramichi (Maine and Tetagouche localities) terranes are placed in association with the western part of the PVA-TEB north of West Avalonia according to the present geographical distribution of major tectonic elements in the Canadian and adjacent New England Appalachians by van Staal *et al.* (2009). These terranes may have experienced significant lateral displacement during their docking history with Laurentia and their inferred mutual position within the western PVA-TEB is uncertain. The resolution of the faunal data is not high enough to make these adjustments; however, they do support that the terranes were closely related as they share a substantial amount of genera. The early mid Darriwilian Bellewstown terrane is of very low diversity (3 genera) and difficult to compare with the other Darriwilian localities. It contains one key genus of the Celtic Province (*Ahtiella*), which it shared with the Anglesey and Central Newfoundland terranes and the Precordillera. It is placed within the PVA-TEB together with the other Celtic terranes but in close proximity to the Leinster Terrane as it amalgamated with the northeastern boundary of this during the Salinic Orogeny (Murphy *et al.* 1991). The Grangegeeth terrane was not investigated and is, hence, not included in the reconstructions. Pollock *et al.* (2012) described a peri-Gondwanan faunal affinity for this terrane in the Early Ordovician and Harper & Parkes (1989) and Murphy *et al.* (1991) specified a strong faunal affinity towards Laurentia in the early Sandbian indicating the Grangegeeth Terrane was part of the PVA-TEB during the Early–Late Ordovician.

Longitudinal separation

The Ganderian terranes contain a significant proportion of taxa that originated in the Baltic Palaeobasin and migrated to the terranes during the Darriwilian and their longitudinal separation from Baltica is indicated as being reasonably limited in the reconstructions. East Avalonia was likely attached to the trailing edge of Ganderia as explained earlier and the distinct brachiopod assemblages from this area cannot be used as an indicator of longitudinal separation but are rather a function of ecology and oceanic current directions.

Latitudinal drift of the Ganderian–Avalonian segment

The pronounced latitudinal distance between the Central Newfoundland Terrane and East Avalonia of 31°S and over 50°S at 470 Ma, respectively, described by Van der Voo *et al.* (1991), may partly be explained by the presence of the trailing edge of

Ganderia and the widening Tetagouche-Exploits backarc basin in between these sites; and partly by a NW–SE orientation of the long axis of the Ganderian–Avalonian segment following van Staal *et al.* (2012) (Figs 6 and 7). Some earlier papers separated a selection of the west Ganderian terranes from the leading edge of Avalonia and placed them in association with an isolated volcanic arc far north in the Iapetus Ocean based on the palaeomagnetic data, also applied herein, as well as faunal signals (e.g. Harper *et al.* 1996, 2013; Mac Niocaill *et al.* 1997). The new reconstructions suggested in the present work accommodate the palaeomagnetic and faunal signals as well as the substantial amount of evidence now relating the Leinster and Monian Composite terranes to Ganderia.

There are some uncertainties regarding the latitudinal positions of the palaeocontinents and terranes, however. For one, the width of the trailing Ganderia margin has not been discussed previously and is merely based on the descriptions of Ganderia by van Staal *et al.* (2012) describing it as a very narrow microcontinent. The width of the leading Ganderia margin corresponds to the terrane widths indicated in the unmodified BugPlates maps (Torsvik 2014). Adjusting the widths of the leading and trailing Ganderia margins will of course change the calculated latitudes somewhat. Following Van der Voo *et al.*'s (1991) palaeomagnetic results for Summerford on the Central Newfoundland Terrane (Ganderia) at 470 Ma, the palaeomagnetic results of Trench & Torsvik (1992) for the Welsh Basin (East Avalonia) in the latest Darriwilian (at 459 Ma) and the drift rates for the leading edge of Ganderia and Avalonia and spreading rates for the Tetagouche-Exploits backarc basin from van Staal *et al.* (2012), the latitudinal positions for these localities are calculated for the mid Darriwilian (463 Ma) and early Sandbian (457 Ma) time-slices. This was carried out by calculating the drift distances from the drift rates and measuring the corresponding latitudinal drift using Google Earth. Thus, a spreading rate of 4 cm/a results in a ca. 480-km-wide Tetagouche-Exploits backarc basin in the mid Darriwilian and a 720-km-wide basin in the early Sandbian.

Assuming a latitudinal position of the Summerford locality of about 37°S for 470 Ma (which is within the 31°S ±8° of Van der Voo *et al.* 1991), a drift rate of ca. 9 cm/a positions this site at 32°S in the mid Darriwilian and 27°S in the early Sandbian, corresponding to northward drift distances of about 630 and 540 km, respectively. If 31°S is used as a starting point, this results in a far closer position between the PVA-TEB terranes and Laurentia in the early Sandbian than can be explained by the faunal signals.

Likewise, a position of 45°S ± an undefined uncertainty for the Welsh Basin in the latest Darriwilian places this site at about 48°S in the mid Darriwilian. This position requires a backarc basin ~800 km wide, which does not correspond to the calculated width of 480 km. Applying the calculated width of the basin to a latitude of 34–35°S for the distal margin of the leading Ganderian edge in the 463-Ma reconstruction (see Fig. 6), the trailing Ganderia margin across from the Summerford locality is situated at about 39°S. This positions the central part of the Welsh Basin at ~44°S in the current reconstruction. An early Sandbian Tetagouche-Exploits backarc basin width of ~720 km likewise places the trailing Ganderia margin opposite Summerford at ~36°S and the central part of the Welsh Basin at 41–42°S (Fig. 7). This corresponds to a drift rate of ca. 5 cm/a as specified by van Staal *et al.* (2012) for 460 Ma and a northward drift distance for Avalonia and the trailing Ganderia margin of ca. 300 km.

The NW–SE orientation of the Ganderian–Avalonian segment was accordingly adjusted to accommodate the calculated latitudes for Summerford and the Welsh Basin.

The Laurentian margin and the Scottish terranes are positioned at about 15°S with the Laurentian Red Indian Lake arc at 19°S in the mid Darriwilian and early Sandbian reconstructions following Torsvik *et al.* (1990, 1996) and van Staal *et al.* (2012). The western Popelogan-Victoria arc terranes (Central Newfoundland and Miramichi terranes) docked with the terranes of the Red Indian Lake arc by the Laurentian margin during the Taconic Orogeny 3 around 460–450 Ma corresponding to the late Darriwilian – early Katian (van Staal *et al.* 2009). If the timing of the Taconic Orogeny 3 by van Staal *et al.* (2009) is followed this would imply that the western Popelogan-Victoria arc may have docked with the peri-Laurentian arc contemporarily with the faunas of the current study, placing the eastern PVA-TEB and Avalonia in very close proximity to Laurentia. The faunal evidence herein does not support such a close position to Laurentia for any of the analysed time-slices. The south Iapetus Ocean and Tornquist Sea localities shared several brachiopod taxa with the north Iapetus localities in the Darriwilian and even more in the early Sandbian but they also maintained a significant number of genera endemic to the south Iapetus-bordering sites as well as taxa otherwise only occurring in high-latitude peri-Gondwana. Hence, the arc–arc collision most likely happened later than the early Sandbian, and the minimum collision age of 450 Ma is here assumed. With an assumed continued drift rate of 9 cm/a for the leading edge of

Ganderia, the 13 million years between the arc–arc collision and the average age of the mid Darriwilian time-slice (463 Ma) position the western Popelogan-Victoria arc approximately 1170 km south of the Red Indian Lake arc, corresponding to a latitude of 30°S, which is 2° north of the assumed latitudinal position of the Summerford locality. In the early Sandbian, the western Popelogan-Victoria arc would thus have been in a position 540 km closer to the Red Indian Lake arc at a distance of approximately 630 km to this, corresponding to latitudinal position between 24 and 25°S.

Brachiopod provinces

The distribution of the Darriwilian – early Sandbian rhynchonelliformean brachiopod genera within the Iapetus Ocean, Rheic Ocean and the Tornquist Sea are analysed using multivariate techniques to resolve the biogeographical affinities of the Leinster Terrane. Faunal data from the Tramore and Dunabrattin Limestone formations were analysed using a recently updated version of the early mid Darriwilian – early Sandbian part of the comprehensive worldwide Ordovician database of rhynchonelliformean brachiopod genera used by Harper *et al.* (2013). The database was compiled by the first author of this study and C.M.Ø. Rasmussen and supplemented with data by the other co-authors of the 2013 paper. The database is locality-based and grouped into two time-slices, covering the early mid–latest Darriwilian and the early Sandbian respectively, for which the data have been converted into presence–absence matrices. The biogeographical results are mapped using modified maps from BugPlates (basemaps from Torsvik 2014) in accordance with the new Iapetus reconstructions. A locality list including biostratigraphical dating and inferred palaeobathymetry (Depth Range zones) is provided in Appendix A.

One of the challenges of applying such large databases is the wide range of biofacies, diversities and time spans represented by the numerous sampling localities often resulting in a range of confusing and sometimes different biogeographical signals even within a single basin. For the larger, more complex continental units, data have been combined from the large number of localities available in the database in order to provide a more average biogeographical signal to simplify identification of brachiopod provinces. Provinces are defined on the basis of their endemic taxa (essentially singletons); however, in grouping together faunas into possible provinces, these taxa are eliminated from the

matrices as explained below. The numbers of endemic genera are discussed under each province.

The brachiopod faunas were analysed using the multivariate cluster and principal coordinates (PCO) analysis. It is common procedure to remove rare occurrences from the presence–absence matrices as they may introduce random permutations in a given data set, and removal of the rare taxa shifts the mean and allows for a statistically meaningful interpretation regarding how different any two samples are (Maples & Archer 1988). Singletons (essentially endemic taxa) and localities containing less than three genera were accordingly removed from the matrices before analysis. Cosmopolitan taxa were likewise removed as they occurred within a very wide range of biofacies and latitudes making them unfit for biogeographical analysis. The few deepwater localities were included in the cluster and PCO analyses. The Simpson coefficient (C/N_1) was chosen for the cluster and PCO analyses as it is among the most accepted coefficients for biogeographical analysis and is suitable for analysing 'sparse data' (Maples & Archer 1988; Rasmussen 1998). The assemblages in the data matrices display a high variation in diversity between the different localities and the matrices also contain far more 0s than 1s ('sparse data'). The Simpson coefficient is normalized to the less diverse fauna (Rasmussen 1998) and produces reliable results as long as the number of variables (0 and 1) is higher than 100 (Maples & Archer 1988). One challenge of applying the cluster analysis is that the clustering procedure will produce a dendrogram no matter how little the assemblages are related. The 'strength' of the clusters can only be judged by investigating the clustering levels within the dendrogram. The principal coordinates (PCO) analysis is an ordination method and does not have this problem as it does not enforce any hierarchy upon the data set. It is important, though, to compare the results of the cluster and PCO analysis with the actual faunas and biogeographically important taxa as the presence of foreign genera in a locality may be related to periods of global sea-level highstands and/or long-distance dispersal of brachiopod larvae by the global oceanic current systems rather than geographical closeness.

Faunal distribution

Several factors determine the distribution of rhynchonelliform brachiopod taxa, such as the nature of their larvae. Planktotrophic larvae may have dominated within the Obolellata, Strophomenata, Protorthida and Orthida, while lecithotrophic larvae appear to have been common within the Pentamerida, Rhynchonellida, Atrypida, Athyrida and

Spiriferida (Freeman & Lundelius 2005). Plank-totrophic larvae feed in the water column and might stay there for some time before settling on a sub-strate (Freeman & Lundelius 2005), while lecitho-trophic larvae do not feed in the water column and presumably settle within a few days (Neuman & Harper 1992; Richardson 1997; Harper *et al.* 2013). Hence, dispersal of taxa with lecithotrophic larvae would normally only be for short distances per gen-eration without the help of a transporting medium, for example pumice (see below), while taxa with planktotrophic larvae would be transported for longer distances before settlement, although, not far enough to cross an ocean (L.E. Popov personal com-munication 2016). The direction of larvae dispersal is largely determined by oceanic current flow for both types of larvae as they are small and not very good swimmers. Survival of the larvae and establish-ment of viable brachiopod populations on the ocean floor or submerged terranes is likely determined by substrate type and availability, food supply, competi-tion, water depth and temperature, water clarity, energy, oxygen levels, etc. Richardson (1997) emphasized the role of substrate type as being the most important factor controlling the brachiopods' capability of colonization in an area and noted that no direct evidence exists to show that depth, temper-ature, latitude and energy of the water limit bra-chiopod distribution. Even so, biogeographical and brachiopod provinciality studies over the last dec-ades seem to confirm such dependence at least among certain non-generalist taxa in the past. Trans-port distance of each generation of brachiopod lar-vae (past and present) is basically a function of current velocity and duration of the larval stage, although transport distance can be greatly enhanced

if the larvae settle on floating objects such as pumice.

Oceanic wind-driven surface currents were the primary factor determining brachiopod dispersal in the Darriwilian, when the majority of taxa were asso-ciated with shelfal biofacies. Successful dispersal of the shelf-bound taxa across vast oceans such as the Iapetus could have taken place by 'island hopping' as suggested by Owen & Parkes (2000) and/or by pumice rafting as suggested by Neuman (1984) and described in a study of recent marine taxa by Bryan *et al.* (2012). 'Island hopping' might have been facil-itated by volcanic hotspots in the oceans producing intraoceanic chains of volcanic islands for bra-chiopod larvae to colonize. Pumice rafting is also a likely possibility owing to the explosive nature of subduction related volcanism which was extensive in the Ordovician. Bryan *et al.* (2012) observed pumice rafting for a number of marine invertebrate groups following the 2006/2007 eruption of the Home Reef volcano in Tonga. They discovered that pumice raft-ing is a very rapid and effective way of dispersing biomass with more than 80 species that underwent an over 5000 km journey in only 7–8 months. Of these two scenarios, pumice rafting would have been the fastest way of dispersal involving as little as one generation depending on brachiopod life cycle rate, while 'island hopping' probably took a significant amount of brachiopod generations to cross a wide ocean.

Details on the oceanic distribution of the bra-chiopod genera shared between the analysed regions are given in Tables 5–7 following the defined distri-butional categories in Table 4. Two categories may be combined in the tables (e.g. IS + T) if a genus was recorded from two of the defined regions.

Table 4. Brachiopod distribution categories of Tables 5–7.

Symbol	Category	Definition
C	Cosmopolitan	Genus distribution across a wide span of latitudes and commonly also longitudes
CP	Celtic Province	Genus confined to the Celtic Province in the early mid–latest Darriwilian
E	Endemic	Only recorded from a given area
I	Iapetus Ocean	Genus widespread within the Iapetus Ocean (East Avalonia, PVA-TEB terranes, Laurentian Platform, peri-Laurentian arc terranes, Precordillera)
IN	North Iapetus Ocean	Genus confined to the north Iapetus Ocean within the Low-latitude province (Laurentian Platform, peri-Laurentian terranes)
IS	South Iapetus Ocean	Genus confined to the southern Iapetus Ocean (East Avalonia, PVA-TEB terranes) but not unique to the Celtic province
LlP	Low-latitude Province	Genus more or less widely distributed within the Low-latitude Province (e.g. Laurentian Platform, peri-Laurentian terranes, Siberia, Kazakh terranes, South China, low-latitude East Gondwana)
Rh	Rheic Ocean	Genus confined to the Rheic Ocean (high-latitude peri-Gondwana and Avalonia)
T	Tornquist Sea	Genus confined to the Tornquist Sea of the Baltic palaeobasin
W	Widespread	Genus recorded from a number of localities with a wide geographical, non-global distribution
x	–	Pre-early mid Darriwilian distribution not investigated

Avalonia is regarded as being part of both the northernmost Rheic Ocean and the south Iapetus Ocean making up the boundary between these. East Avalonia includes a mix of taxa that originated in the Darriwilian of high-latitude peri-Gondwana and low-latitude Laurentia, respectively, and subsequently migrated to East Avalonia. Hence, genera only recorded from high-latitude peri-Gondwana and East Avalonia are defined as associated with the Rheic Ocean, and taxa restricted to the PVA-TEB terranes and East Avalonia are related to the south Iapetus Ocean.

Darriwilian provinces and faunas

The investigated Darriwilian faunas were part of the key Dapingian–Darriwilian interval for brachiopod diversification and continental disparity. Forty matrix localities, many of them compiled from several sampling localities, and 201 genera have been identified globally, of which 26 sites and 160 genera were recorded from the Iapetus Ocean, Rheic Ocean and Tornquist Sea in this study. North Estonia, the Scottish Midland Valley Terrane and the eastern and southeastern Laurentian platform have very high diversities of 32–40 genera; five sites have high diversities of 20–25 genera, while a number of sites are of moderate to low diversity with 7 and 10 sites having diversities of 10–18 and 1–9 genera, respectively. Eighty-three genera occur at only one of the compiled matrix localities, whereas *Sulevorthis sensu* Jaanusson & Bassett (1993) occurs at 11 sites, *Paralenorthis* at nine sites and *Hesperorthis*, *Sowerbyella* (*Sowerbyella*) and *Valcourea* at eight sites. The matrix of localities primarily covers the early mid to latest Darriwilian, although some localities include stratigraphy overlapping with the earliest parts of the Darriwilian as well as the early Sandbian, as it was not possible to separate the brachiopod occurrences into more precise biostratigraphical intervals.

The five Dapingian–Darriwilian brachiopod provinces discussed by Harper *et al.* (2013) are recognized in the Darriwilian part of this investigation, that is the Baltic, Celtic, High-latitude, Low-latitude and Toquima-Table Head provinces. Additionally, a late Darriwilian fauna characterized by Scoto-Appalachian taxa was recognized from the Scottish Midland Valley Terrane making this the earliest occurrence of the Scoto-Appalachian fauna.

The Toquima-Table Head and Celtic provinces were originally erected because their brachiopod assemblages could not be readily accommodated within the Low-latitude (previously Laurentian), Baltic and High-latitude (previously Mediterranean)

provinces (Harper *et al.* 1996). They include distinct faunal assemblages commonly, but not always, associated with volcanic arc–backarc environments developed on the continental margins of Laurentia and the leading edge of Ganderia, respectively. The Toquima-Table Head Province was situated within the low-latitude, warm-water environments that surrounded the trans-equatorial Laurentian Platform (Harper *et al.* 1996), whereas those of the Celtic assemblage were affected by somewhat cooler waters within the low mid-latitudes.

The Low-latitude Province on the Laurentian Platform

From the Tremadocian until the late Katian, the Low-latitude Province extended across a low-latitudinal belt centred on the Equator from the Laurentian Platform in the west to Australia and Tasmania in the east (Harper *et al.* 2013). The Low-latitude Province was characterized by brachiopod genera widely distributed within the province, transitional faunas between regions within the province, as well as endemics associated with their respective platforms, basins and terranes.

The Low-latitude Province on the shallow-water Laurentian carbonate platform is clearly identified in the cluster and PCO analyses (Figs 8 and 9, respectively). The platform faunas (shallow- and deep-water) are of low to very high diversity with a total diversity for the platform of 51 genera and a total endemicity of 35%. The Laurentian Platform operated as a species pump indicated by the presence of a significant proportion of Dapingian–Darriwilian Laurentian taxa in the PVA-TEB terranes, East Avalonia and Baltica. The term 'species pump' was introduced by Harper *et al.* (2013, p. 129) and describes 'regions with great species diversity which acted as centres that allowed extra-regional spread of taxa' resulting in increased γ-diversity.

Non-cosmopolitan brachiopod taxa commonly co-occurring within sites of the mid–latest Darriwilian Laurentian Platform and some of the marginal terranes include *Ancistorhyncha*, *Atelelasma*, *Colaptomena*, *Dactylogonia*, *Dorytreta*, *Glyptorthis*, *Hesperorthis*, *Isophragma*, *Mimella*, *Multicostella*, *Oepikina*, *Onychoplecia*, *Ptychopleurella*, *Rostricellula*, *Sowerbyella* (*Sowerbyella*), *Sphenotreta*, *Stenocamara*, *Sulevorthis*, *Titanambonites* and *Valcourea*. Of these, *Ancistorhyncha*, *Dorytreta*, *Sphenotreta*, *Stenocamara* and *Titanambonites* are not recorded outside the platform are regarded as endemic during this period. *Multicostella*, *Onychoplecia*, *Sulevorthis* and *Valcourea* are also recorded from one or more peri-Laurentian terranes with the

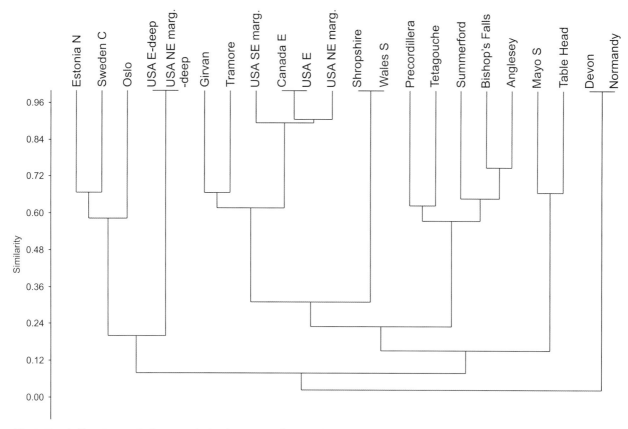

Fig. 8. Darriwilian (463 Ma) cluster analysis, Simpson similarity index. Low-latitude Province on the Laurentian Platform: Canada E, USA E, USA E-deep, USA NE marg., USA NE marg.-deep, USA SE marg. Toquima-Table Head Province: Mayo S, Table Head. Scoto-Appalachian fauna: Girvan. Celtic Province: Anglesey, Bishop's Falls, Precordillera, Summerford, Tetagouche. Mixed fauna – Leinster Terrane: Tramore. Mixed fauna – East Avalonia: Shropshire, Wales S. High-latitude Province: Devon, Normandy. For explanation of localities, see Appendix A.

two latter furthermore occurring in some PVA-TEB terranes (Table 5). *Isophragma, Mimella, Oepikina* and *Rostricellula* were apparently restricted to the Laurentian–peri-Laurentian–Siberian–Gorny Altaian transect of the Low-latitude Province.

The deep-water localities of the Laurentian Platform appear to be closely related to Baltica from the cluster and PCO analyses (Figs 8 and 9, respectively), although they share only one genus, *Leptelloidea* (Table 5). The deep-water siliciclastic facies share more genera with the shallow-water sites of the Laurentian carbonate platform and are regarded as most closely related to these.

An early Scoto-Appalachian fauna

The late Darriwilian Scottish Midland Valley Terrane (Girvan) is of very high diversity containing 40 different genera with an endemicity of about 25%. It contains almost all brachiopod taxa typical of the Sandbian Scoto-Appalachian Province

including *Bimuria, Christiania, Colaptomena, Dactylogonia, Isophragma, Leptellina, Paurorthis, Phragmorthis, Plectorhis, Taphrorthis* and *Valcourea* (see Harper 1992; Parkes 1992; Candela & Harper 2014). It contains only few genera characteristic of the Dapingian–Darriwilian Toquima-Table Head assemblages, that is *Eremotoechia, Leptellina, Phragmorthis* and *Valcourea*, of which only *Eremotoechia* was not also a common constituent of the Scoto-Appalachian Province (see Harper 1992 and Neuman & Harper 1992). This shows that the Scottish Midland Valley Terrane developed a Scoto-Appalachian signature prior to the rest of the marginal Laurentian sites otherwise associated with the Toquima-Table Head Province (Fig. 10).

The cluster and PCO analyses (Figs 8 and 9, respectively) indicate strong faunal links between the Scottish Midland Valley Terrane, the Leinster Terrane and the Laurentian Platform. The limestones and calcareous siliciclastics of the Midland Valley Terrane share nine non-cosmopolitan taxa

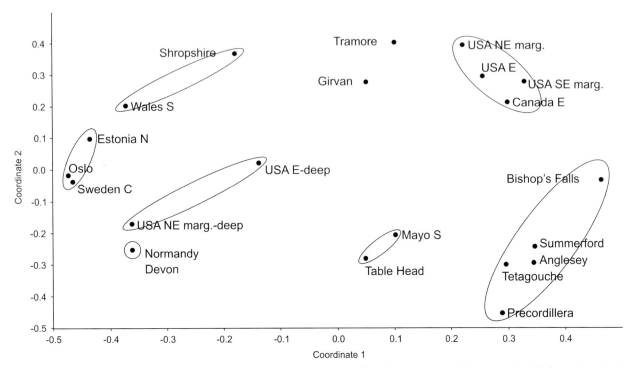

Fig. 9. Darriwilian (463 Ma) PCO analysis, Simpson similarity index. Low-latitude Province on the Laurentian Platform: Canada E, USA E, USA E-deep, USA NE marg., USA NE marg.-deep, USA SE marg. Toquima-Table Head Province: Mayo S, Table Head. Scoto-Appalachian fauna: Girvan. Celtic Province: Anglesey, Bishop's Falls, Precordillera, Summerford, Tetagouche. Mixed fauna – Leinster Terrane: Tramore. Mixed fauna – East Avalonia: Shropshire, Wales S. High-latitude Province: Devon, Normandy. For explanation of localities, see Appendix A.

with both sites, of which five (*Colaptomena, Dactylogonia, Sowerbyella* (*Sowerbyella*), *Sulevorthis, Valcourea*) are recorded from all three areas (Table 5). It likewise shares five non-cosmopolitan genera with the peri-Laurentian Toquima-Table Head Province (*Camerella, Eremotoechia, Nothorthis, Phragmorthis, Valcourea*) indicating strong faunal links; however, only *Eremotoechia* commonly co-occurs within this province. Taxa that possibly migrated from Laurentian marginal and intraoceanic sites to the Leinster Terrane were already established in East Avalonia and the Celtic PVA-TEB terranes before deposition of the Tramore Limestone commenced, which is discussed in later sections.

The Toquima-Table Head province

The Toquima-Table Head assemblage was originally defined by Ross & Ingham (1970) based on occurrences in carbonate rocks on the Laurentian Platform margin of western Nevada (Toquima Range) and the western Newfoundland foreland basin site of Table Head (Maletz & Egenhoff 2011). It also occurred in a number of Dapingian–Darriwilian localities in siliciclastic and volcaniclastic facies on arc-related peri-Laurentian terranes, as well as on

the Siberian Platform margin and the Chu-Ili Terrane of Kazakhstan (Neuman & Harper 1992). The late Darriwilian Toquima-Table Head faunas are recognized from the shallow-water siliciclastic and volcaniclastic facies of the South Mayo Terrane and the shallow-water carbonate facies of west Newfoundland (Table Head Group). Together, the two localities constituting the late Darriwilian province have a high diversity of 27 genera and a high endemicity of 48%, although local diversity and endemicity is lower.

Typical to variably occurring within the province are the genera *Aporthophyla, Eremotoechia, Idiostrophia, Leptella, Leptellina, Liricamera, Neostrophia, Onychoplecia, Orthidiella, Orthidium, Phragmorthis, Polytoechia, Pomatotrema, Protoskenidioides, Rhysostrophia, Stenocamara, Syndielasma, Syntrophina, Taphrodonta, Trematorthis, Toquimia* and *Trondorthis* (Harper 1992; Neuman & Harper 1992; Harper *et al.* 2013). Commonly co-occurring taxa in the late Darriwilian Toquima-Table Head localities include *Archaeorthis, Idiostrophia, Leptella, Orthidium, Pleurorthis, Rhysostrophia, Toquimia* and *Trondorthis*, of which, for example, *Leptella, Pleurorthis* and *Rhysostrophia* are not recorded outside the province and are regarded as endemic during the late Darriwilian. Although they do not co-occur in the two late

Table 5. Darriwilian (463 Ma); distribution of taxa shared between the faunally distinct regions. See Table 4 for distribution categories.

	HG	Av	LT	An	Be	WPT	Prec	Balt	TTH	MV	Laur-Deep	Laur
Ahtiella				CP	CP	CP	CP					
*Atelelasma**				W								W
Bimuria				W						W		W
Camerella				W		W	W		W	W		W
*Christiania**		W						W		W		
*Colaptomena**		I	I							I		I
Corineorthis	Rh	Rh										
Cyrtonotella										IN		IN
Dactylogonia		I		I						I		I
Eodalmanella	(W)					(W)						
Eremotoechia									IN	IN		IN
Ffynnonia				CP			CP					
Glyptomena										W		?
Glyptorthis		W	W									W
Hesperonomia				I					I			
Hesperorthis	?	C	C					C		C		C
Howellites	Rh + IS		Rh + IS									
Idiostrophia†							E		E			
Inversella						IS + T		IS + T				
*Isophragma**										LlP		LlP
Jaanussonites					CP	CP						
Kullervo		IS + T?						?				
Laticrura										IN		IN
Leptellina†*	C		C				C	C		C		C
Leptelloidea								IN + T			IN + T	
Mimella										LlP		LlP
Monorthis				CP		CP	CP					
Multicostella										IN		IN
Palaeoneumania				CP		CP						
Nothorthis									W	W		
Onychoplecia†									IN			IN
Orthidium†							W		W			
Oxoplecia		I + T						I + T		I + T		
Palaeostroph.								IN + T		IN + T		IN + T
Paralenorthis				C	C	C	C		C			C
*Paurorthis**	C		C	C		C	?			C		C
Petroria							LlP	LlP				
Phragmorthis†							W		W	W		W
Platystrophia		IS + T						IS + T				
Plectorthis				?		W				W		W
*Productorthis**				C		C	C					C
Ptychopleurella						LlP + CP				LlP + CP		LlP + CP
Reinversella				CP			CP					
Rostricellula		W								W		W
Rugostrophia				I		I	I		I			
Salopia	?	Rh + IS	Rh + IS									
Sivorthis								IN + T		IN + T		IN + T
Skenidioides				W		W	W			W		W
Sowerbyella (S.)		W	W					W		W		W
Sulevorthis			I	I		I				I	I	I
Taphrorthis										IN		IN
Tissintia	Rh	Rh										
Titanambonites											IN	IN
Triplesia		W								W	W	W
Tritoechia				W		W	W		W			
Ukoa						IS + T		IS + T				
Valcourea†*		I				I				I		I

Regions and localities: An, Anglesey; Av, East Avalonia (Shropshire, South Wales); Balt, Baltica (Estonia N, Oslo, Sweden C); Be, Bellewstown; HG, high-latitude peri-Gondwana (Bohemia, Devon, Normandy, Spain); Laur, Laurentian Platform (Canada E, USA E, USA NE marg., USA SE marg.); Laur-deep, Laurentian Platform deep-water sites (USA E-deep, USA NE marg.-deep); LT, Leinster Terrane (Tramore); MV, Scottish Midland Valley Terrane (Girvan); Prec, Precordillera; TTH, Toquima-Table Head sites (South Mayo, Table Head); WPT, western PVA-TEB (Bishop's Falls, Indian Bay, Maine, Summerford, Tetagouche). *Palaeostroph.*, *Palaeostrophomena*. *Scoto-Appalachian taxa. †Toquima-Table Head taxa.

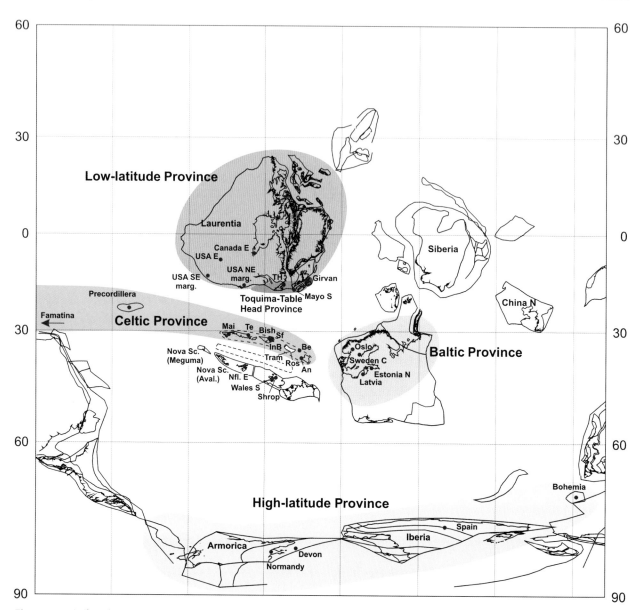

Fig. 10. Darriwilian (463 Ma) provinces. The current study mainly focused on brachiopod provinciality within the Iapetus and Rheic oceans and the Tornquist Sea and, hence, the interpreted provinces are only shown for the investigated areas. The full extent of the provinces (excluding the Toquima-Table Head and Scoto-Appalachian provinces) is given in Harper *et al.* (2013). Note that the Celtic Province was present on all the Ganderian terranes north of Avalonia except the Leinster Terrane. Laurentian Platform localities: Canada E, east Canada; USA E, east USA; USA NE marg., northeast margin of USA; USA SE marg., southeast margin of USA. Peri-Laurentian terrane localities: Gi, Girvan (Scottish Midland Valley Terrane); May, South Mayo Terrane; TH, Table Head (west Newfoundland foreland basin). Precordilleran localities: Precordilleran platform. PVA-TEB terrane localities: An, Anglesey (Monian Composite Terrane); Be, Bellewstown Terrane; Bish, Bishop's Falls (Central Newfoundland Terrane); InB, Indian Bay (Central Newfoundland Terrane); Mai, Maine (Miramichi Terrane); Ros, Rosslare (Monian Composite Terrane); Sf, Summerford (Central Newfoundland Terrane); Te, Tetagouche (Miramichi Terrane); Tram, Tramore (Leinster Terrane). East Avalonian localities: Shrop, Shropshire; Wales S, South Wales. Baltic Platform localities: Estonia N, north Estonia; Latvia, Latvia undiff.; Oslo, Oslo region, Norway; Sweden C, central Sweden. Peri-Gondwanan localities: Bohemia, Devon, Normandy, Spain. Other: China N, North China; China S, South China; Nfl E, East Newfoundland terrane; Nova Sc, Nova Scotia.

Darriwilian localities, *Eremotoechia* and *Syntrophina* are likewise recorded from South Mayo, and *Aporthophyla* and *Onychoplecia* from Table Head.

The late Darriwilian Toquima-Table Head fauna is closely related to the Laurentian Platform and the Scottish Midland Valley Terrane as described above, and it shares six non-cosmopolitan genera with both

sites, of which four (*Camerella*, *Eremotoechia*, *Phragmorthis*, *Valcourea*) are recorded from all three sites (Table 5). It likewise shares five non-cosmopolitan genera with the Celtic PVA-TEB terranes (*Camerella*, *Hesperonomia*, *Rugostrophia*, *Tritoechia*, *Valcourea*), although they maintain their distinctness as well-defined provinces in the cluster and PCO analyses

(Figs 8 and 9, respectively), a trend that was likewise recognized by, for example, Neuman & Harper (1992), Harper (1992) and Harper *et al.* (1996). Some genera shared between the Toquima-Table Head and Celtic provinces were commonly associated with peri-continental arc terranes and microplates in the early mid–late Darriwilian. These are *Rugostrophia*, reported from the north and south Iapetus arc terranes and the Precordillera, and *Camerella* and *Tritoechia* that commonly occurred in arc terranes, although the former is also reported from the Laurentian Platform and the latter from the south Chinese platform within the Low-latitude Province.

The mixed faunas of the Leinster Terrane and East Avalonia

The brachiopod faunas recorded from the Ganderian PVA-TEB terranes and East Avalonia (Shropshire and South Wales) all contain a significant number of taxa that possibly migrated from Laurentian marginal and intraoceanic sites, although the PVA-TEB terranes, excluding the Leinster Terrane, are assigned to the distinct Celtic Province. Only a few of the taxa associated with Laurentia and its margins during the later Mid Ordovician originated in the Laurentian region but rather they originated in other areas of the world such as South China, Siberia, East Gondwana, etc. from where they migrated via the Laurentian sites to the south Iapetus–Tornquist region.

The Leinster Terrane shows strong affinities with the Scottish Midland Valley Terrane in the cluster and PCO analyses, with East Avalonia being distantly related to both (Figs 8 and 9, respectively). The carbonate-rich siliciclastic, volcaniclastic and argillaceous limestone facies of the Leinster Terrane contains a mixed fauna originating from several places around the globe such as high-latitude peri-Gondwana, the mid-latitude Baltic Palaeobasin and the low-latitude Laurentian region. The Darriwilian Leinster fauna contains few of the characteristic taxa from the late Darriwilian Toquima-Table Head Province (*Leptellina*, *Valcourea*) and more from the Scoto-Appalachian Province (*Colaptomena*, *Leptellina*, *Paurorthis*, *Valcourea*), all of which were present on the Scottish Midland Valley Terrane. The shallow to mid/outer shelf siliciclastic facies of East Avalonia likewise includes fauna of peri-Gondwanan origin, taxa that possibly migrated from the Laurentian region as well as a few genera commonly occurring in the Scoto-Appalachian Province (*Christiania*, *Colaptomena*). The Leinster Terrane and East Avalonia do not contain any taxa diagnostic of the Celtic

Province (see below) and they are not readily assigned to any of the currently defined provinces due to their mixed faunas.

The Leinster Terrane

The brachiopod fauna of the Leinster Terrane is of moderate diversity (14 taxa) with possibly two endemic genera. The new porambonitid genus *Hibernobonites* described from both the Tramore and Dunabrattin Limestone formations has not been recorded elsewhere in the Darriwilian. The porambonitid genera currently assigned to '*Porambonites*' *sensu lato* are in much need of revision, however, and new studies of these and allied genera may reveal other occurrences of *Hibernobonites*. Earlier in the Ordovician (Floian–Dapingian) *Hibernobonites* possibly occurred on the South Mayo Terrane marginal to Laurentia as described in the systematics section. Significantly *Hibernobonites* may have migrated to higher latitudes during the Katian, appearing in the Mediterranean Province during the Boda Warming Event as part of the *Nicolella* Fauna (see Colmenar 2015 and the systematics section herein). Another possibly endemic genus is the oldest recorded taxon closely resembling *Tetraphalerella*; however, it is extremely rare and only represented by one ventral valve in the Tramore Limestone Formation. The Tramore Limestone contains five non-cosmopolitan taxa that possibly migrated from Laurentian marginal and intraoceanic sites to the Leinster Terrane during the later Darriwilian, that is *Colaptomena* (Cooper 1956), *Dactylogonia* (Hofmann 1963), *Sowerbyella* (*Sowerbyella*) (Cooper 1956), *Sulevorthis* (Cooper 1956) and *Valcourea* (Neuman & Harper 1992; unpublished database). It furthermore includes *Salopia* of probable mid-latitude West Gondwanan origin (Cordillera Oriental, Benedetto 2003b; Fortey & Cocks 2003), *Platystrophia* of Baltic origin (Hansen & Harper 2003) and *Howellites* of high-latitude peri-Gondwanan origin (Iberia, Villas 1985). Some of these genera are recorded from several sites pre-dating and contemporary with the Tramore Limestone; for example, *Sulevorthis* and *Valcourea* were already present on the Celtic PVA-TEB terranes prior to deposition of the Tramore Limestone as were *Colaptomena* and *Sowerbyella* (*Sowerbyella*) in East Avalonia. Parkes (1992) noted in his description of the Duncannon Group faunas that migration of taxa with a marginal Laurentian (Scoto-Appalachian), Baltic and Gondwanan origin to the Leinster Terrane had taken place earlier in the Ordovician.

At the generic level, only *Howellites*, *Platystrophia* and *Salopia* may be valuable biogeographical indicators as they had a restricted distribution. *Howellites* is reported from the High-latitude Province (Villas 1985) and the Leinster Terrane; *Salopia* is recorded from the Leinster Terrane and East Avalonia (Lockley & Williams 1981) and possibly the High-latitude Province and mid-latitude West Gondwana within the Rheic Ocean (Cocks & Lockley 1981; Benedetto 2003b; respectively); and *Platystrophia* is reported from the Leinster Terrane and the Baltic Province (Zuykov & Harper 2007). *Howellites* later acquired a virtually cosmopolitan distribution during the Sandbian–early Katian (Cocks & Torsvik 2002).

The Tramore Limestone Formation shares some of its Darriwilian species with other Iapetus-bordering sites in which these species occurrences pre-dates the Tramore Limestone. *Sulevorthis* aff. *S. blountensis* from Tramore is very similar to *S. blountensis* from the Laurentian Platform (Cooper 1956) and *Valcourea confinis* possibly had its first occurrence on the Scottish Midland Valley Terrane (e.g. Williams 1962), although it is contemporary with the Tramore occurrence and may have originated on the Leinster Terrane. Likewise, *Sowerbyella* (*Sowerbyella*) *antiqua* occurred in East Avalonia (Lockley & Williams 1981) prior to deposition of the Tramore Limestone. *Paurorthis parva* was present on the Baltic Platform in the Dapingian (e.g. Egerquist 2003) but disappeared already in the earliest Darriwilian (L.E. Popov personal communication 2016). Hence, a descendant of this likely migrated from Baltica to the Leinster Terrane resulting in the presence of the morphologically very similar *Paurorthis* aff. *P. parva* recorded from the Tramore Limestone.

East Avalonia
The East Avalonian fauna is of high diversity containing 22 different genera with an endemicity of 32%. It includes five non-cosmopolitan taxa that possibly migrated from the Laurentian marginal and intraoceanic sites which were all described by Cooper (1956): *Christiania*, *Colaptomena*, *Oxoplecia*, *Sowerbyella* (*Sowerbyella*) and *Triplesia*, as well as *Rostricellula* that possibly originated in Svalbard (Hansen & Holmer 2010), although the material is incompletely and poorly preserved, or Laurentia (Hofmann 1963) and later became near-cosmopolitan during the Sandbian–early Katian (Cocks & Torsvik 2002). *Oxoplecia* and *Triplesia* had a restricted distribution and are additionally only recorded from Baltica (Hints & Rõõmusoks 1997) and Alaska (Rasmussen *et al.* 2012), respectively. East Avalonia furthermore includes two genera, *Corineorthis* and *Tissintia*, which are characteristic of the High-latitude Province in the later part of the Darriwilian, but possibly originated in low-latitude South China in the Dapingian (Xu & Liu 1984), and

the mid-latitude West Gondwanan Cordillera Oriental in the earliest Darriwilian, respectively (Benedetto 2003b; Fortey & Cocks 2003). The limited distribution of *Oxoplecia*, *Triplesia*, *Corineorthis*, *Salopia* and *Tissintia* makes them useful biogeographical indicators. No genera originating within the Baltic Palaeobasin or the PVA-TEB terranes are reported from East Avalonia.

The Celtic province

The concept of the Celtic Province has been subject to much debate (e.g. Neuman & Harper 1992; Boucot 1993; Cocks & McKerrow 1993; Harper *et al.* 1996, 2009, 2013; Williams *et al.* 1996). The province was originally defined by Williams (1973, p. 249) with the Dapingian–lower Darriwilian Treiorwerth Formation from Anglesey constituting the reference assemblage for the Celtic Province. Cocks & McKerrow (1993) argued against the definition of the province, but, for example, Neuman & Harper (1992) and Harper *et al.* (1996, 2009, 2013) and this study tested this biogeographical unit using new data and modern multivariate statistical methods and its integrity remains consistently reproducible. The early mid–latest Darriwilian faunas of the present investigation are compared with some of the Dapingian–early Darriwilian faunas in Table 6, including the Treiorwerth Formation from Anglesey. The latter were not included in the cluster and PCO analyses as the objective of these was to give a high-resolution overview of the palaeobiogeography contemporary with the Tramore and Dunabrattin Limestone formations.

The Celtic Province had its widest extent during the Dapingian–early Darriwilian, embracing a disparate group of continental fragments and arc terranes extending across the globe within the lower to mid-latitudes. The Darriwilian Celtic faunas are recorded from the Precordillera and most of the investigated terranes of the PVA-TEB, that is Miramichi (Maine, Tetagouche), Central Newfoundland (Bishop's Falls, Indian Bay, Summerford), Bellewstown and Monian Composite terranes (Anglesey, Rosslare) (see Table 6). During this period, the Precordillera constituted a carbonate platform setting, while the PVA-TEB terranes included a variety of siliciclastic, volcaniclastic and carbonate facies in the volcanic arc–backarc setting. The current analyses show that the mid–late Darriwilian East Avalonia and Leinster Terrane do not include any of the taxa diagnostic of the Celtic Province and share very few genera with the province at all (Table 6). East Avalonia only shares one near-cosmopolitan genus (*Skenidioides*) with the other Celtic sites and the Leinster Terrane shares three non-cosmopolitan

Table 6. Genera shared between the shelf localities of the PVA-TEB terranes, Avalonia and Argentina and their relation to the Celtic Fauna.

	Fam	Prec	Mai	Teta	Sum	Bish	InB	Be	Ros	AnT	An	Tra	Shr	Wa
Ahtiella		(CP)			(CP)	(CP)	(CP)	(CP)			(CP)			
Antigonambonites				(CP)	(CP)									
Camerella	x	W			W									
Colaptomena												I	I	I
Dactylogonia											I	I		
Dalmanella													W	W
Ffynnonia	CP	CP							CP		CP			
Glyptorthis												W	W	W
Hesperonomia	x								x		I			
Hesperonomiella	C									C				
Hesperorthis												C		C
Horderleyella													W	W
Jaanussonites			CP		CP									
Orthidium		I					x							
Oxoplecia													I	I
Monorthis	CP	CP		CP						CP	CP			
Palaeoneumania					(CP)						(CP)			
Paralenorthis	x	C		C	C	C		C	x	x	C			
Paurorthis					C					?	C	C		
Platytoechia			CP				CP							
Productorthis	x	C	x	C	C	C			x	x	C			
Rectotrophia									CP	CP	CP			
Reinversella		CP							CP		CP			
Rhynchorthis									CP	CP	CP			
Rugostrophia	x	I		I	I					x	I			
Salopia												Rh+IS		Rh+IS
Skenidioides	x	W		W			x		x	x	W			W
Sowerbyella (S.)												W	W	W
Sulevorthis	x		x	I	I	I	x			x	I	I		
Tissintia													Rh	Rh
Treioria									CP	CP	CP			
Tritoechia	x	W	x	W	W		x			x	W			
Valcourea						I					I			

Localities: An, Anglesey (Monian Composite Terrane); AnT, Anglesey-Treiorwerth Fm (Monian Composite Terrane); Be, Bellewstown Terrane; Bish, Bishop Falls (Central Newfoundland Terrane); Fam, Famatina; InB, Indian Bay (Central Newfoundland Terrane; Mai, Maine (Miramichi Terrane); Prec, Precordillera; Ros, Rosslare (Monian Composite Terrane); Shr, Shropshire (East Avalonia); Sum, Summerford (Central Newfoundland Terrane); Teta, Tetagouche Group (Miramichi Terrane); Tra, Tramore Limestone Fm (Leinster–Lakesman Terrane); Wa, South Wales (East Avalonia). Taxonomical definition of *Sulevorthis* follows Jaanusson & Bassett (1993). Grey columns represent Dapingian–early Darriwilian faunas that are not included in the multivariate analyses. For distribution categories, see Table 4.

(*Dactylogonia, Sulevorthis, Valcourea*); they are thus excluded from the Celtic Province (Fig. 10). The new Darriwilian palaeogeographical reconstruction, which includes Anglesey and Rosslare in the PVA-TEB, combined with the palaeogeographical positions of the Dapingian–early Darriwilian Celtic-dominated Otta (Norway), north Chinese and Argentinian sites of Neuman & Harper (1992) and Harper *et al.* (2013), limits the total latitudinal extent of the province to ~10–36°S. The investigations of, for example, Neuman & Harper (1992) and Harper *et al.* (1996, 2013) associated the present PVA-TEB terranes with Avalonia at the higher southern latitudes and, hence, described a wider latitudinal extent for the Celtic Province between 10°S and 50°S.

The Celtic faunas are characterized by a large number of endemic brachiopod taxa, some cosmopolitan and some widespread forms as well as genera at the beginning or end of their stratigraphical ranges (Harper *et al.* 2013). Genera occurring in two or more Celtic sites within the total extent of the province through space and time are *Ahtiella, Famatinorthis, Ffynnonia, Jaanussonites, Monorthis, Palaeoneumania, Paralenorthis, Platytoechia, Productorthis, Rectotrophia, Reinversella, Rhynchorthis, Rugostrophia, Skenidioides, 'Orthambonites'* (possibly *Sulevorthis*), *Treioria* and *Tritoechia* (Harper 1992; Harper *et al.* 2013; this study). Genera commonly co-occurring in, and endemic to, the early mid–late Darriwilian Celtic sites include *Ahtiella, Antigonambonites, Ffynnonia, Jaanussonites, Monorthis, Palaeoneumania, Platytoechia, Rectotrophia, Reinversella, Rhynchorthis* and *Treioria*. These are marked as 'CP' in Table 6 and referred to as the 'Celtic diagnostic genera'. Except for *Ahtiella, Antigonambonites* and *Palaeoneumania*, all of these seem to have had their first occurrence within the sites associated with the Celtic Province (i.e. the Precordillera: *Reinversella* (Herrera & Benedetto 1991), *Monorthis* (Benedetto 2003b). Famatina: *Ffynnonia* (Benedetto 2003b). Maine: *Jaanussonites*, *Platytoechia* (Neuman 1984). Rosslare: *Rectotrophia*, *Rhynchorthis*, *Treioria* (Neuman & Harper 1992; unpublished database)). *Ahtiella* may have originated at either the Precordillera (Herrera & Benedetto 1991) or the Baltic Platform (Rasmussen 2005; Rasmussen & Harper 2008) as the recorded occurrences for these sites are contemporary in the early Darriwilian. *Antigonambonites* and *Palaeoneumania* presumably originated in the late Floian of the Baltic Province (Egerquist 2003; Hansen & Harper 2005). Less common endemics of Celtic origin that are only recorded from the western PVA-TEB terranes include *Acanthorthis* (Zhan & Jin 2005), *Guttasella* (Zhan & Jin 2005), *Fistulogonites* (Neuman 1984) and *Dolerorthis* (Neuman 1984). All of these, except *Dolerorthis*, became extinct by the end Darriwilian. *Dolerorthis* later acquired a cosmopolitan distribution and persisted into the Early Devonian (Williams & Harper 2000).

The proportion of non-cosmopolitan taxa of Baltic origin within the Celtic PVA-TEB terranes is significant and, besides *Antigonambonites*, *Palaeoneumania* and possibly *Ahtiella*, includes, for example, *Ingria* (Egerquist 2003), *Inversella* (Hansen & Harper 2003) and *Ukoa* (Rasmussen 2007). Fortey & Cocks (2003) suggested that *Atelelasma* may have originated in the Baltic Palaeobasin. However, the present study, as well as the *Treatise on Invertebrate Paleontology* (Williams & Harper 2000), does not record *Atelelasma* from Baltica. According to L.E. Popov (personal communication 2016), there is no record of *Atelelasma* in Baltoscandia, but rather it may have evolved in an intraoceanic island setting from an early clitambonitoid that migrated from Baltica.

The Celtic faunas in the PVA-TEB terranes also include several taxa of Laurentian marginal and intraoceanic origin such as *Christiania* (i.e. Neuman 1984), *Paralenorthis* (e.g. Bates 1968), *Paurorthis* (Neuman 1984) and *Valcourea* (Bruton & Harper 1981; Neuman 1984). No taxa of East Avalonian origin were recorded from the Celtic PVA-TEB terranes.

Explanation of the pronounced difference in faunal content between the Leinster Terrane and the Celtic PVA-TEB terranes is not straightforward. Their faunal distinctness may, at least in part, be accounted for by timing of the faunas. All the Celtic assemblages originated prior to deposition of the Tramore Limestone, leaving time for development of the specialized endemic Celtic diagnostic genera before the Leinster Terrane became suitable for colonization. The specialized Celtic genera were likely outcompeted by opportunistic generalist taxa in the race for colonization of the Leinster Terrane, providing an explanation to why they have not been reported there.

Some of the Celtic diagnostic taxa survived until the early–mid Sandbian in central Newfoundland and Anglesey, which comprised a museum for these faunas. The last occurrence of Celtic diagnostic genera includes *Jaanussonites*, *Monorthis* and *Rectotrophia* which were recorded by Bates (1968) from the upper *gracilis*–lower *foliaceous* zones in Anglesey.

The Baltic Province

The Baltic Province was associated with the Baltic Platform and included a unique fauna consistently distinguished by multivariate analyses applying different similarity indices (e.g. Harper *et al.* 2009, 2013; this study). Based on Rhynchonelliformean brachiopod evidence, the province has been identified from as early as the Tremadocian (Harper *et al.* 2013) and it persisted until the end of the Darriwilian, confirmed by Harper *et al.* (2013) and this study.

The biofacies differentiation of the Ordovician faunas in the East Baltic, including that of the brachiopods, was first demonstrated by Männil (1966), who also established the main facies belts within the Baltic Palaeobasin (Hints & Harper 2003). Later accounts on brachiopod biofacies differentiation within this area were given by, for example, Jaanusson (1973, 1976) and Hints & Harper (2003). The current study includes faunas from the North Estonian (Estonia N), Central Baltoscandian (Sweden C) and the Oslo Confacies belts (see Nielsen 2004). The North Estonian faunas were characterized by high diversity (34 genera) and high endemicity (65%) and were associated with a shallow-water, upper ramp setting dominated by more or less argillaceous carbonate deposits (see Hints & Harper 2003). The Swedish and Norwegian faunas were of significantly lower diversity (4 and 6 genera, respectively) with significantly less or no endemic fauna (0% and 33%, respectively) and occupied the deeper-water, lower ramp environments, which were less suitable for habitation by brachiopods (Hints & Harper 2003), and included a variety of carbonate and siliciclastic facies.

The distinctly differentiated biofacies makes it difficult to give a generalized characterization of the brachiopod faunas inhabiting the Baltic Province. Taxa recorded from more than one of the investigated localities are rare and include *Alwynella*, *Christiania*, *Leptelloidea*, *Leptestia* and *Nicolella*, of which *Alwynella* was endemic to the province.

The Baltic Province is well defined in the cluster and PCO analyses (Figs 8 and 9, respectively) due to its unique faunas, but it still shares some taxa with the other sites predominantly of Laurentian and south Chinese origin. It shares the highest number of non-cosmopolitan genera with the Scottish Midland Valley Terrane (six) and a maximum of five genera with East Avalonia (Table 5), with *Kullervo* being questionably identified from Baltica. The Baltic Province likewise shares three taxa (*Inversella*, *Platystrophia*, *Ukoa*) with the PVA-TEB terranes, all of which were apparently of Baltic origin. None of the genera recorded from the Baltic Palaeobasin originated from the PVA-TEB and only *Kullervo* possibly originated in East Avalonia.

The High-latitude Province

The High-latitude Province is well defined in the cluster and PCO analyses (Figs 8 and 9, respectively) and is characterized by cold-water brachiopod faunas of low to very low diversity (3–9 genera) obtained from shallow to mid shelf siliciclastic facies and oolitic ironstones from the peri-Gondwanan microplates of Armorica (Devon, Normandy), Iberia (Spain) and Bohemia. Total diversity for the Darriwilian province includes 15 genera with a total endemicity of 40%. Spain and Bohemia contain very few non-endemics and non-cosmopolitan taxa and faunas were of such low diversity after removal of these taxa from the matrices that they were not appropriate for statistical analysis.

Taxa commonly co-occurring within the province include *Corineorthis*, *Tafilaltia* and *Tissintia*, of which *Tafilaltia* was endemic. *Corineorthis*, of high-latitude origin and *Tissintia* of mid-latitude West Gondwanan origin have not been recorded outside the Rheic Ocean in the early mid to latest Darriwilian, but were able to exist at both the northern (East Avalonia) and southern (high-latitude peri-Gondwana) extremities of this ocean.

Darriwilian faunal migration patterns: implications on oceanic current configuration

Understanding the forcing mechanisms of the modern surface current systems is essential for describing ancient current systems, assuming that the physical principles controlling today's surface currents were also operational in the past. The modern oceanic surface currents are primarily driven by winds and, due to rotation of the Earth, follow directions that are based on fundamental physical principles, for example the Ekman transport related to the Coriolis force (Kaiser *et al.* 2005; Huang 2009). Wind-driven circulation generally refers to the circulation in the upper ocean (depth <1 km) which is primarily driven by wind stress (Huang 2009). The large-scale oceanic surface currents, averaged over the whole of the water column, move perpendicular to the wind due to interaction with the Coriolis force (Kaiser *et al.* 2005). Movement is to the right of the wind in the Northern Hemisphere and to the left in the Southern Hemisphere (i.e. Ekman transport) (Kaiser *et al.* 2005). Consequently, the modern oceanic surface currents form large circular patterns called gyres. Anticyclonic gyres move in a clockwise direction in the Northern Hemisphere and anticlockwise in the Southern Hemisphere, while cyclonic gyres move in the opposite direction on their respective hemispheres (Kaiser *et al.* 2005; Huang 2009). Large obstacles in the pathway of a given current, for example continent coastlines and islands, may deflect the currents according to the shape of the obstacle as shown by Klinger (1993).

The suggested Darriwilian surface currents, plotted on the mid Darriwilian palaeobiogeographical reconstruction (Fig. 11A, B), are based on the detailed, state-of-the-art, mid Ordovician synthetic ocean surface circulation models of Pohl *et al.* (2016a) which were modified in this investigation by the inferred brachiopod migrational patterns and also somewhat simplified to only show the suggested currents of Pohl *et al.* (2016a) relevant for the present discussions. Brachiopod migration was studied by tracking biogeographically important taxa through space and time using an updated version of the Ordovician brachiopod database applied by Harper *et al.* (2013). Pohl *et al.* (2016a) constructed their models using the coupled ocean-atmosphere general circulation model FOAM combined with the palaeogeographical reconstructions by Torsvik & Cocks (2009). The present work adds Ganderia to the palaeogeographical reconstructions. Pohl *et al.* (2016a) furthermore investigated the sensitivity of circulation pattern to the atmospheric CO_2 content, which, however, remains uncertain due to a lack of Ordovician proxy data (see Pohl *et al.* 2016a and references therein). This resulted in three Mid Ordovician models for three atmospheric CO_2 levels (4 PAL, 8 PAL and 16 PAL). They concluded that circulation patterns were more or less stable at high CO_2 levels between 8 and 16 PAL while modelling at the lowest CO_2 level, 4 PAL, changed circulation patterns somewhat, although the major currents remained the same. The brachiopod migrational patterns reported by this study can be explained by the low-CO_2-level model (4 PAL) for the early Darriwilian (Fig. 11A) and the high-CO_2-level model

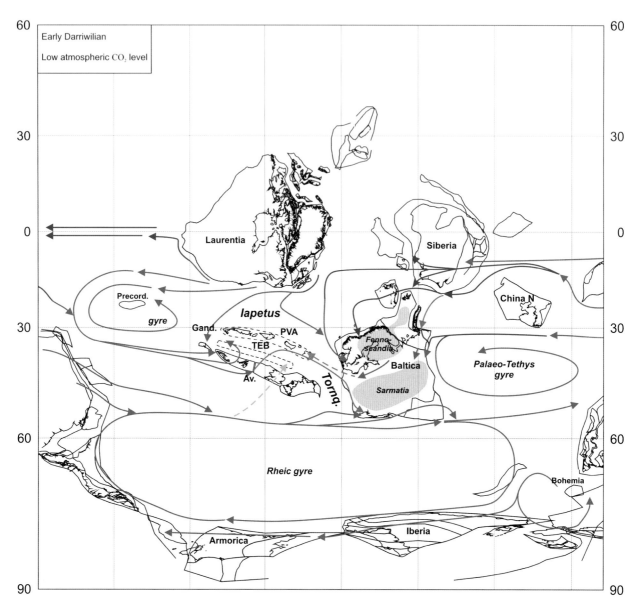

Fig. 11. **A** (above), Early Darriwilian currents projected on the 463-Ma palaeobiogeographical reconstruction. Red and blue arrows represent warm and cool–cold currents, respectively, and were redrawn from Pohl *et al.* (2016a). Solid grey arrows represent modifications of the currents reconstructed by Pohl *et al.* (2016a) and do not indicate temperature. The dashed dark green arrow represents currents flowing from the Baltic Palaeobasin to Ganderia. The dashed light green arrow represents currents flowing from the Rheic gyre to Avalonia/Ganderia. Brown transparent areas indicate the emerged areas of Sarmatia and Fennoscandia and were redrawn from Pohl *et al.* (2016a). Av., Avalonia; China N, North China; Gand., Ganderia; Precord., Precordillera; PVA, Popelogan-Victoria arc; TEB, Tetagouche-Exploits backarc basin. **B** (p. 37), mid–late Darriwilian currents projected on the 463-Ma palaeobiogeographical reconstruction. Red and blue arrows represent warm and cool–cold currents, respectively, and were redrawn from Pohl *et al.* (2016a). Solid grey arrows represent modifications of the currents reconstructed by Pohl *et al.* (2016a) and do not indicate temperature. The dashed dark green arrows represent currents flowing from the Baltic Palaeobasin to the PVA-TEB. The dashed light green arrow represents currents flowing from the Rheic gyre to Avalonia/Ganderia. Brown transparent areas indicate the emerged areas of Sarmatia and Fennoscandia and were redrawn from Pohl *et al.* (2016a). Av., Avalonia; China N, North China; Gand., Ganderia; Precord., Precordillera; PVA, Popelogan-Victoria arc; TEB, Tetagouche-Exploits backarc basin.

(8–16 PAL) for the mid–late Darriwilian (Fig. 11B). This is in accordance with Darriwilian temperature changes described by, for example, Rasmussen *et al.* (2016) and Pohl *et al.* (2016b). Early Darriwilian temperatures were cool (corresponding to low atmospheric CO_2 levels), associated with an ice age,

encouraging the initial phases of the Great Ordovician Biodiversification Event among the benthos (Rasmussen *et al.* 2016 and references therein). This, however, may have been relatively short-lived as temperatures recovered slightly, later in the Darriwilian (Pohl *et al.* 2016b) (corresponding to higher

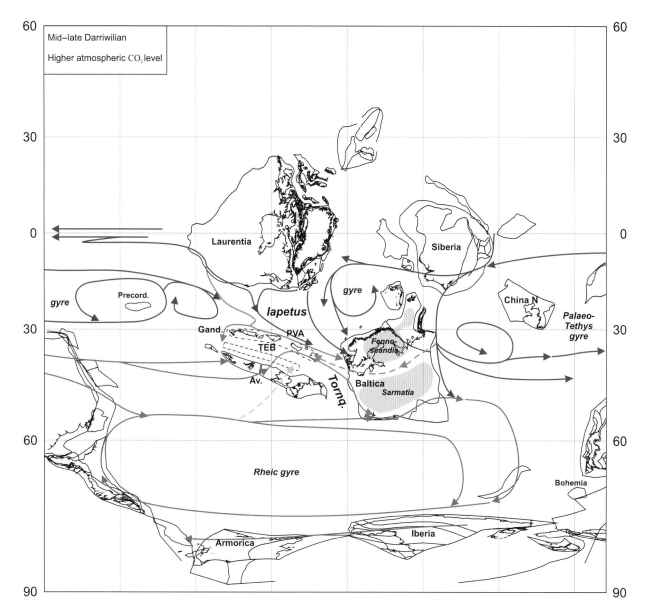

atmospheric CO$_2$ levels). Note that the reconstruction used for the low-CO$_2$-level model (Fig. 11A) was not palaeogeographically adjusted for the early Darriwilian. Ganderia/Avalonia would have occupied a slightly more southerly position than shown on the figure and the Tetagouche-Exploits backarc basin (TEB) would have been slightly narrower. These minor changes would likely not have influenced the greater flow patterns of the oceanic currents, though.

The models of Pohl *et al.* (2016a) do not indicate any currents that may account for the reported brachiopod migrations from the Baltic Palaeobasin to the Ganderia and from high-latitude Gondwana to East Avalonia/Ganderia. As it is not possible to precisely determine when the brachiopods migrated, suggested currents responsible for brachiopod dispersal are shown as green dashed arrows in both the early and mid–late Darriwilian models (Fig. 11A, B, respectively). It is highly likely that additional fossil assemblages are yet to be described adding new information to the known fossil record and the inferred migration patterns and taxa originations. Hence, these interpretations should only be considered as a guideline and additional data are needed for a more comprehensive evaluation. In the models suggested by this work, the red and blue arrows were redrawn from Pohl *et al.* (2016a) and represent warm and cool–cold currents, respectively. For details about Ordovician sea surface temperature modelling, see Pohl *et al.* (2014). The solid grey arrows represent modifications of the currents reconstructed by Pohl *et al.* (2016a) and do not indicate temperature.

Migrations from the Laurentian margins and intraoceanic islands to the south Iapetus–Tornquist region

The Mid Ordovician was an interval of considerable
provincial complexity. Much of the faunal data are
incomplete, not least that from key intraoceanic
environments, now either abducted and amalgamated in mountain belts or lost during subduction
processes. In addition, the most recent migrational
routes may not always indicate the origin of a given
taxon, only its last journey. Faunal migration into
the south Iapetus–Tornquist region appears to have
taken place in three stages: The first phase was
marked by an early Darriwilian colonization by a
few Laurentian marginal and intraoceanic genera of
the PVA-TEB terranes and included forms such as
Christiania from Central Newfoundland (Neuman
1984), *Paralenorthis* from Anglesey and elsewhere in
the Celtic Province (e.g. Bates 1968), *Paurorthis*
from Central Newfoundland (Neuman 1984) and
Valcourea from central Newfoundland and Otta
(Bruton & Harper 1981; Neuman 1984). The second mid Darriwilian phase included, for example,
the arrival of *Christiania* and *Sowerbyella* (*Sowerbyella*) in East Avalonia and Baltica (Hints &
Rõõmusoks 1997) and *Colaptomena* in East Avalonia. Nevertheless, current evidence suggests that
eastern Gondwana may have been the primary origin of *Colaptomena* and related taxa, taking advan-
tage of westward flowing currents for dispersion
towards Laurentia and the Iapetus terranes. The
third, late Darriwilian phase, was possibly responsible for migration of *S. blountensis* (which experienced some morphological changes prior to its
occurrence in the Tramore Limestone) and *V. confinis* to the Leinster Terrane, although they could
have resided on the other Ganderian terranes prior
to their occurrence in the Tramore Limestone. The
third phase was also responsible for migration of
Oxoplecia to East Avalonia and Baltica (Hints &
Rõõmusoks 1997), *Rostricellula* and *Triplesia* to East
Avalonia and *Palaeostrophomena* to Baltica (Hints
& Rõõmusoks 1997). These events may be correlated probably with global drowning events and
sea-level highstands in the Darriwilian observed
from the Baltic sections by Nielsen (2004). This
pattern is supported by the discussions by Fortey &
Cocks (1988) on faunal distribution related to
transgressive–regressive cycles. In periods of transgression and sea-level highstand, oceanic biofacies
are brought shelf-wards facilitating a wider dispersal
of marine taxa.

The currents responsible for transporting brachiopod larvae from the East and northern Iapetus
Ocean to the south Iapetus–Tornquist region are
most easily explained by the high-CO_2-level model
(Fig. 11B) as these currents are predicted to flow
directly from the Far East, past Siberia and from the
entire palaeosouthern margin of Laurentia to all of
the PVA-TEB terranes. However, the minor migration reported from the early Darriwilian might be
explained by a combination of the minor drowning
event and the suggested currents of the low-CO_2-
level model that would only have facilitated indirect
transport of brachiopod larvae from Laurentia to the
southern Iapetus Ocean. The two more extensive
migrational events reported from the mid and late
Darriwilian, respectively, was possibly caused by a
combination of the more pronounced sea-level
drownings and a change in oceanic flow (Fig. 11B)
caused by increased atmospheric CO_2 levels and
likely other yet unknown factors as well.

The distance between the palaeosouthern Laurentian margin and Ganderia was likely extensive as
suggested in the Darriwilian palaeobiogeographical
reconstruction (Fig. 6); for example, the distance
between the Leinster and Midland Valley terranes
may have been around 2500 km. Dispersal of shallow-water taxa between the North and South Iapetus
Ocean possibly happened both by 'island hopping'
and 'pumice rafting' as described in the 'Faunal distribution' section. The presence of a volcanic hotspot in the Darriwilian Iapetus Ocean is a likely
scenario (P.M. Holm personal communication
2015). The volcanic hotspot would have produced a
chain of volcanic islands extending at least for some
distance across the Iapetus Ocean providing sub-
strates for possible brachiopod settlement (see Neuman 1984; Bruton & Harper 1985). Other types of
islands might also have been present in the Iapetus
Ocean facilitating 'island hopping'. 'Pumice rafting'
is also relevant owing to the subduction related volcanism which took place in the PVA and RIL on
either side of the Iapetus Ocean (Neuman 1984).
Volcanic activity in the PVA and RIL must have produced significant amounts of pumice in periods of
extensive volcanism.

The current flowing from West Gondwana
towards the PVA-TEB suggested by both the high-
and low-CO_2-level models (Fig. 11A, B, respectively)
may account for the distribution of Argentinian taxa
in the Celtic Province (e.g. the PVA-TEB terranes),
the Leinster Terrane and East Avalonia.

Migrations within the south Iapetus–Tornquist
region

The faunal evidence shows that dispersal of brachiopod taxa within the south Iapetus–Tornquist
region took place in several directions.

Faunal migration from East Avalonia to the Leinster Terrane most likely took place directly across the Tetagouche-Exploits backarc basin (TEB). They share two widespread genera, *Colaptomena* and *Glyptorthis*, and the species *Sowerbyella* (*Sowerbyella*) *antiqua*, all of which occurred in the mid Darriwilian of East Avalonia prior to deposition of the Tramore Limestone. Faunal migration from East Avalonia to the Leinster Terrane may have been facilitated by the currents suggested to have flowed eastwards from West Gondwana and along the mid-latitudes to be deflected northwards across Avalonia. Pohl *et al.* (2016a) did not include Ganderia in their reconstructions, and hence, we have modified their model and extended the northward flowing current across the TEB to the Popelogan-Victoria arc (PVA) where it is deflected towards the east (Fig. 11A, B). The eastward deflection is in accordance with both the high- and low-CO_2-level models of Pohl *et al.* (2016a). North-to-south transport of larvae may also have existed, inferred by the occurrence of the later Darriwilian taxa of presumed Laurentian origin in East Avalonia. North-to-south dispersal of taxa might have been facilitated by minor currents or eddies but their flow directions are highly speculative and, hence, it was not attempted to reconstruct them in Figure 11A,B. Taxa recorded from the Leinster Terrane that were not observed from East Avalonia (see Table 5) may for some reason not have been able to settle there. Furthermore, East Avalonia and the PVA-TEB terranes only shared one near-cosmopolitan genus (*Skenidioides*) (see Table 6) and the lack of faunal exchange between the PVA-TEB terranes and East Avalonia might suggest that their faunas had reached equilibrium and/or their respective ecological settings were unsuitable for colonization of taxa from the opposing area. The later migrations of Laurentian taxa to East Avalonia may indicate temporary formation of new ecospace due to the inferred drowning events, but it does not explain why Leinster taxa did not migrate south to East Avalonia and why the new Laurentian taxa did not settle on any of the PVA-TEB terranes during these migrations.

Faunal migration between the closely positioned terranes of the PVA-TEB may have taken place in both a west to east and east to west direction indicated by the non-cosmopolitan, non-Celtic genera shared between the Leinster Terrane and the PVA-TEB terranes, that is *Dactylogonia* (Anglesey, e.g. Bates 1968), *Sulevorthis* and *Valcourea*. These were present on various terranes of the PVA-TEB prior to deposition of the Tramore Limestone.

The presence of a significant number of Baltic genera and some species in the PVA-TEB terranes indicates that migration from Baltica probably took place directly between these two areas facilitated by surface currents flowing westwards from the Baltic Palaeobasin (here named the Baltic current for simplicity) as suggested with the dark green dashed arrows in Figure 11A, B. The Baltic current is not supported by any of the models by Pohl *et al.* (2016a) making brachiopod dispersal from Baltica to Ganderia hard to explain without modifying their models. A possible explanation to the presence of Baltic fauna in Ganderia involves a westward flowing current from South China (flowing between north China and Siberia in Fig. 11A, B) that branched on the eastern margin of the Baltic Platform to run towards Ganderia between the emerged areas of Fennoscandia and Sarmatia (brown transparent areas on Fig. 11A, B; redrawn after Pohl *et al.* 2016a). South China was possibly positioned on low southerly latitudes east of North China (Torsvik & Cocks 2013) but a more exact position was not investigated further in the present paper. This scenario is supported by the presence of taxa that originated in South China earlier in the Ordovician but these could also have been dispersed by other currents continuing along the northern margin of Baltica (see Fig. 11A, B). The low-CO_2-level model by Pohl *et al.* (2016a) indicates a Palaeo-Tethyan current flowing westwards between Sarmatia and Fennoscandia, although it is deflected southwards when encountering the Avalonian current (corresponding to the Ganderian current in this study) (see Fig. 11A). It is still unclear what facilitated the migration of the rhynchonelliformean brachiopods from the Baltic Palaeobasin to Ganderia and this subject is in need of further investigation.

No mid–late Darriwilian taxa of Baltic origin have been reported from East Avalonia in the present investigation. *Ingria* and *Athiella* were present in Avalonia in the earliest Darriwilian (latest Volkhov; see Hints & Harper 2003), but they have not been reported from the later parts of the Darriwilian in the present study. As described above none of the Baltic (or otherwise) genera already present in mid–late Darriwilian Ganderia managed to colonize the Anglo-Welsh Basin during this period, while taxa presumably migrating from Laurentia settled there indicating southward dispersal across Ganderia possibly took place. Both the low- and high-CO_2-level models (Fig. 11A, B, respectively) indicate that, through the Darriwilian, brachiopod dispersal from the Baltic Palaeobasin to Avalonia might have been hampered by the suggested current (modified from Pohl *et al.* 2016a) flowing from the PVA to the western margin of Baltica and south along this. Another explanation of the lack of younger Darriwilian Baltic taxa in East Avalonia might also be a two-step, 'minor' extinction event in the early Llanvirn (at the base and top of the Aseri Stage, respectively), which could have eliminated the genera that would otherwise have migrated directly from Baltica to Avalonia (L.E. Popov personal communication 2016).

The reported absence of Anglo-Welsh taxa in the Baltic Palaeobasin can be explained by the late Darriwilian lowstand (see Nielsen 2004) as currents are suggested to have flowed from Avalonia to the Baltic Palaeobasin via Ganderia (directly from Avalonia to Baltica in Pohl *et al.* 2016a) in both the low- and high-CO_2-level models (Fig. 11A, B, respectively). Hints & Harper (2003) noted that immigration into the Baltic Palaeobasin from other provinces was associated with major geological events accompanied by changes in sea level, indicating that the Darriwilian drowning events probably were too minor to induce migration. This changed in the early Sandbian when the *gracilis* drowning became substantial enough to facilitate pronounced mixing of brachiopod taxa between the mid-latitudinal palaeoplates (see the section on early Sandbian provinces).

The co-occurrence of Laurentian taxa within the same biozone of East Avalonia and Baltica support the presence of currents diverging from the same point on the palaeosouthern margin of Laurentia as suggested by Pohl *et al.* (2016a) in their high-CO_2-level models (see Fig. 11B). These taxa have not been recorded from the PVA-TEB terranes possibly indicating they were not able to colonize them.

The Rheic gyre and the High-latitude Province

The occurrence of high-latitude taxa in the mid-latitudes of East Avalonia (*Corineorthis*) and the Leinster Terrane (*Howellites*) may be explained by larval transport facilitated by the surface currents of a Rheic gyre (dashed light green arrow, Fig. 11A, B). The Darriwilian Rheic gyre would have been centred on ~60°S, suggesting that it was a cyclonic gyre rotating clockwise, like the modern-day Ross Sea and Weddell Sea gyres by the Antarctic margin (Huang 2009). Clockwise rotation is in accordance with the prevailing wind systems generated by the Coriolis force with the easterlies determining the current directions south of 60°S and the westerlies between 30°S and 60°S and its existence is also supported by both the high- and low-CO_2-level models of Pohl *et al.* (2016a). Servais *et al.* (2014) argued that the Early Ordovician high-latitude cyclonic gyres suggested by Christiansen & Stouge (1999) were contradictory to the Coriolis force. The extent

of the southern gyres suggested by Christiansen & Stouge (1999) should perhaps have been expanded below 60°S and into the easterlies, but with their given palaeoplate positions the existence of such gyres is a possibility.

Darriwilian migration of the mid-latitude West Gondwanan genera (*Tissintia* and *Salopia*) to high-latitude peri-Gondwana may have happened via East Avalonia and/or by currents flowing directly from West Gondwana and into the Rheic gyre (as shown on Fig. 11A, B). It may also have taken place by coastal currents or eddies flowing in the opposite direction of the gyre waters along the peri-Gondwanan margin. However, the exact flow of these is hard to predict and these types of currents were not shown in the models of Pohl *et al.* (2016a). Eddies might have been formed by turbulence as the easterly currents of the Rheic gyre were slowed down and deflected when encountering irregular shorelines along the peri-Gondwanan margin which likely constituted settling grounds for brachiopod larvae most of the way between the mid and high Gondwanan latitudes.

The mixed faunas of the PVA-TEB terranes and East Avalonia support the new palaeogeographical model in that Avalonia and Ganderia occupied a transitional position between the Iapetus and Rheic oceans and the Tornquist Sea. Except for periods of global lowstand, latitudinal distance, current directions and to some extent water temperature were likely the most important barriers obstructing exchange of faunas between the high and low latitudes, but the geographical positions of Ganderia and Avalonia made it possible for genera from both latitudinal regimes to settle and coexist in the mid-latitudes.

The Precordillera

The Precordilleran faunas developed from being dominated by Toquima-Table Head taxa in the early Darriwilian (Herrera & Benedetto 1991; Neuman & Harper 1992) to being dominated by Celtic genera in the early mid–latest Darriwilian (Benedetto 2003b), although they included a significant amount of taxa from both provinces during these periods. Several of the Celtic genera likely originated in and around West Gondwana, and the change from low- to mid-latitudes as well as reduced transport of peri-Laurentian taxa and increased transport of West Gondwanan genera to the Precordillera may account for the observed change in taxa proportions. This is supported by the observations of Benedetto (2003a) who described a progressive change in biogeographical affinities of the Precordilleran faunas from entirely Laurentian to Gondwanan through the Ordovician, with a sudden increase in the proportion of Gondwanan taxa in the late Darriwilian – Sandbian. The change in fauna composition likely reflects the movement of the Precordillera away from the low latitudes and westward flowing currents from the Laurentian margin into the mid-latitudes dominated by eastward flowing currents from West Gondwana. Oceanic flow from Laurentia to the Precordillera in the early Darriwilian is solely supported by the low-CO_2-level model (see Fig. 11A), which is also in accordance with the cool temperatures described for this period by Rasmussen *et al.* (2016). The existence of currents transporting Celtic taxa from West Gondwana to the Precordillera (and further on to the PVA-TEB terranes) is supported by both the low- and high-CO_2-level models (Fig. 11A, B, respectively) indicating that supply of taxa from West Gondwana to Ganderia likely took place through most of the Darriwilian.

Early Sandbian brachiopod provinces

The late Darriwilian and early Sandbian were characterized by a major marine drowning event (Nielsen 2004), the *gracilis* drowning, which caused a significant transgression over the shallow shelfal areas and migration of many deeper-water brachiopod taxa onto the cratons (Harper *et al.* 2013). The Sandbian–early Katian (Caradoc) period also marks the first major migration of benthic shelly communities into deeper-water environments (Sepkoski & Sheehan 1983), reflected in a second peak in global diversification during the Great Ordovician Biodiversification Event (GOBE) (Harper 2006a; Servais *et al.* 2010; Harper *et al.* 2013). These migrations together with the ongoing closure of the Iapetus Ocean were the main causes of the diminished brachiopod provinciality reported between the early mid–latest Darriwilian and early Sandbian. The major migrational event facilitated by the *gracilis* drowning is reflected in the present data by the significant proportion of brachiopod genera that achieved a wider distribution across the Darriwilian – Sandbian transition, for example *Atelelasma*, *Camerella*, *Cyrtonotella*, *Dactylogonia*, *Isophragma*, *Kullervo*, *Ptychopleurella* and *Rostricellula* (see Tables 5 and 7). The present biostratigraphical resolution and availability of early Sandbian fossil localities is not good enough to clarify the migrational patterns of the, geologically speaking, almost instant wider dispersal of taxa caused by this event. It is highly likely, though, that both surface and deep oceanic currents were involved in transport of brachiopod larvae between the now more closely spaced Iapetus-bordering sites. Hence, an attempt to reconstruct early Sandbian oceanic currents in relation to migrational patterns was not undertaken. The number of brachiopod provinces was reduced from five in the Darriwilian to four in the early Sandbian, and early Sandbian provinces identified in the cluster and PCO analyses (Figs 12 and 13, respectively) include the Low-latitude, Scoto-Appalachian, High-latitude and Anglo-Welsh–Baltic provinces (Fig. 14), which were also discussed by Harper *et al.* (2013) in their global Sandbian time-slice.

Thirty-four matrix localities, many of them combined from several sampling localities, and 174 genera have been identified globally, of which 26 localities and 129 genera were recorded from the Iapetus Ocean, Rheic Ocean and Tornquist Sea. There would appear to be a significant drop in diversity between the early mid–latest Darriwilian and early Sandbian time-slices, but it is not significant, as the early Sandbian time-slice covers several million years less than the Darriwilian. Furthermore, Harper *et al.* (2004) recorded an increase in first occurrences compared with last occurrences for all major brachiopod groups across the Darriwilian – Sandbian transition except the Pentamerida. The global diversity increased significantly in the Sandbian to include about 240 genera for the total Sandbian time interval (Harper *et al.* 2013), which in million years largely corresponds to the early mid–latest Darriwilian time-slice of this study. The early Sandbian central Laurentian Platform, Anglesey, Shropshire and North Estonia have high diversities of 21–28 genera, while a number of sites are of lower diversity with 11 and 11 sites having diversities of 10–19 and 1–9 genera, respectively. Fifty-two genera occur at only one site, whereas *Sowerbyella* (*Sowerbyella*) occurs at 10 sites, *Glyptorthis* at nine sites and *Rostricellula* at seven sites. The matrix localities primarily cover the early Sandbian although some localities include stratigraphy overlapping with the latest parts of the Darriwilian and the later parts of the Sandbian, as it was not possible to separate the brachiopod occurrences into more precise biostratigraphical intervals.

The Low-latitude Province on the Laurentian Platform

The Laurentian carbonate platform maintained its distinct faunal signature and is clearly separated from the remainder of the investigated sites in the cluster and PCO analyses (Figs 12 and 13, respectively). The platform faunas (shallow- and deep-water) are of low to high diversity with a total

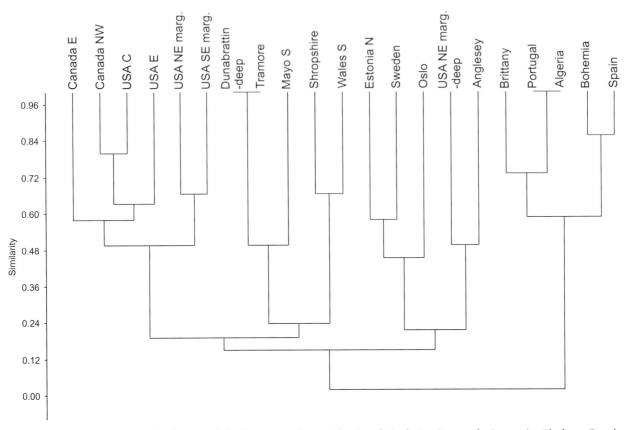

Fig. 12. Early Sandbian (457 Ma). Cluster analysis, Simpson similarity index. Low-latitude Province on the Laurentian Platform: Canada E, Canada NW, USA C, USA E, USA NE marg., USA SE marg., USA NE marg.-deep. Anglo-Welsh–Baltic Province: Anglesey, Dunabrattin-deep, Estonia N, Oslo, Shropshire, Sweden, Tramore, Wales S. Scoto-Appalachian Province: Mayo S. High-latitude Province: Algeria, Bohemia, Brittany, Portugal, Spain. For explanation of localities, see Appendix A.

diversity for the platform of 52 genera and a total endemicity of 46%. While diversity is about the same as in the late Darriwilian (i.e. 51 genera), endemicity is higher (Darriwilian: 35%), indicating continued speciation on a generic level within the Laurentian species pump. Genera that originated on, and were endemic to, the early Sandbian Laurentian Platform include *Catazyga* (Copper 1977), *Cooperea* (Cooper 1956), *Eridorthis* (Cooper 1956), *Salonia* (Cooper 1956) and *Trigrammaria* (Cooper 1956).

Non-cosmopolitan brachiopod taxa commonly co-occurring within sites of the early Sandbian Laurentian Platform include *Chaulistomella, Dactylogonia, Doleroides, Glyptorthis, Mimella, Oepikina, Oxoplecia, Protozyga, Ptychoglyptus, Rostricellula, Skenidioides, Sowerbyella (Sowerbyella)* and *Sowerbyites.* Of these, *Chaulistomella, Doleroides* and *Protozyga* are not recorded outside the Laurentian Platform and are regarded as endemic during this period. *Mimella* and *Sowerbyites* are additionally only reported from the shallow-water South Mayo Terrane.

The marginal Laurentian (USA NE marg.-deep) deep-water fauna share most genera with the Anglo-Welsh–Baltic Province and less taxa with the neighbouring Laurentian and peri-Laurentian sites. The majority of the genera shared with the Anglo-Welsh–Baltic Province were widespread, however, and not good biogeographical indicators. The close affinity between the deep-water Laurentian and shallow-water Anglo-Welsh–Baltic sites was likely caused by deep-water migration between these northern and southern Iapetus-bordering localities and was facilitated by the *gracilis* drowning as well as the closer geographical proximity between these sites. The faunas pre- and post-dating the deep-water fauna were characteristic of the Low-latitude Province, and the deep-water locality on the Laurentian margin is, hence, included in the Low-latitude Province by this study.

The Scoto-Appalachian Province

The typical Scoto-Appalachian taxa described by Harper (1992) and Parkes (1992) were not

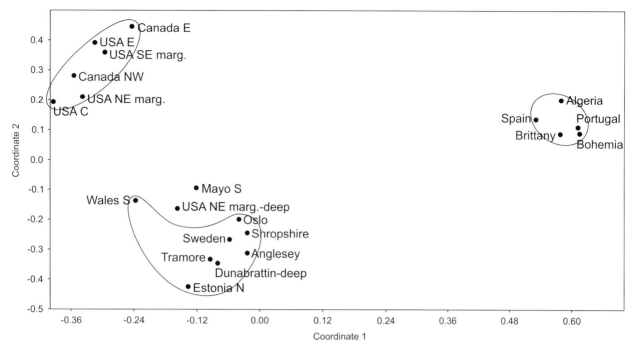

Fig. 13. Early Sandbian (457 Ma). PCO analysis, Simpson similarity index. Low-latitude Province on the Laurentian Platform: Canada E, Canada NW, USA C, USA E, USA NE marg., USA NE marg.-deep, USA SE marg. Anglo-Welsh–Baltic Province: Anglesey, Dunabrattin-deep, Estonia N, Oslo, Shropshire, Sweden, Tramore, Wales S. Scoto-Appalachian Province: Mayo S. High-latitude Province: Algeria, Bohemia, Brittany, Portugal, Spain. For explanation of localities, see Appendix A.

endemic to the province but rather commonly occurring throughout the province, and some are also recorded from other Iapetus-bordering sites (see Table 7). The typical Scoto-Appalachian genera include *Atelelasma*, *Bimuria*, *Christiania*, *Colaptomena*, *Dactylogonia*, *Isophragma*, *Leptellina*, *Paurorthis*, *Phragmorthis*, *Plectorhis*, *Productorthis*, *Taphrorthis*, *Titanambonites* and *Valcourea*. The Scoto-Appalachian Province is identified on the early Sandbian South Mayo Terrane. Sandbian assemblages from the Scottish Midland Valley Terrane (Girvan area, Williams 1962; Harper *et al.* 1985; Cocks & Rong 1989; Candela & Harper 2014) and late Sandbian assemblages from the Irish Midland Valley Terrane (Pomeroy, Candela 2003) include various combinations of the common Scoto-Appalachian taxa.

South Mayo maintained its distinct signature from the Laurentian Platform in the cluster and PCO analyses (Figs 12 and 13, respectively), although it has strong affinities towards the south Iapetus-bordering sites. The South Mayo fauna is of low diversity (6 genera) and contains only *Atelelasma*?, *Isophragma*, *Leptellina*, *Metacamerella*, *Mimella* and 'Porambonites' group (see description of *Hibernobonites* n. gen. for explanation), of which none were endemic. *Atelelasma*, *Isophragma* and *Leptellina* were common constituents of the Scoto-Appalachian Province and *Metacamerella*

and *Mimella* are likewise reported from the Scoto-Appalachian fauna on the Sandbian Scottish Midland Valley Terrane (Girvan area, Williams 1962; Candela & Harper 2014). Although most of these taxa had a more or less widespread distribution within the Iapetus-bordering sites and South Mayo did not contain any endemics, the combination of genera with 83% being associated with the later Scoto-Appalachian faunas links it to this province.

The Leinster Terrane includes a significant Scoto-Appalachian component which was also discussed by Harper (1992) and Parkes (1992). The Tramore Limestone fauna is, however, more closely related to the East Avalonian and Baltic faunas and is described in detail in the section on the Anglo-Welsh–Baltic Province below.

The Anglo-Welsh–Baltic Province

There are strong faunal links between East Avalonia, the PVA-TEB terranes (Leinster Terrane and Anglesey) and Baltica indicated by the cluster and PCO analyses (Figs 12 and 13, respectively) and the tabulated shared genera in Table 7. These transitional faunas form the basis of the Anglo-Welsh–Baltic Province, with province endemics occurring either within restricted areas of the province or throughout the province including the commonly co-occurring

Table 7. Early Sandbian (457 Ma); distribution of taxa shared between the faunally distinct regions. For distribution categories, see Table 4.

	HG	Av	LT	An	Balt	Prec-deep	MaS	Laur-deep	Laur
Atelelasma*			I				?		I
Bicuspina	C	C							
Bilobia				W	W				
Camerella				(W)					(W)
Chonetoidea				W	W				
Christiania*					W				
Colaptomena*		I							I
Cyrtonotella					I + T				I + T
Dactylogonia	W		W						W
Dinorthis				W		W			W
Dolerorthis				W				W	
Drabovia	Rh	Rh							
Eoplectodonta (E.)				(W)	(W)				
Gelidorthis	C	C							
Glyptambonites			I					I	
Glyptorthis		W	W					W	W
Grorudia			IS + T		IS + T				
Harknessella		IS		IS					
Hesperorthis			C		C		C		C
Horderleyella		IS		?					
Howellites	Rh + IS		Rh + IS						
Isophragma*			W				W		
Kullervo				I + T	I + T			I + T	
Leptaena				IS + T	IS + T				
Leptellina*	C		C				C		
Leptestiina			C	C	C				
Metacamerella				I			I		
Mimella							IN		IN
Oepikina					W				W
Onniella	W	W		W	W				
Oxoplecia		I + T			I + T				I + T
Palaeostrophomena				I	?				
Paurorthis*			I		I				I
Platystrophia		IS + T	IS + T	IS + T	IS + T				
Productorthis*			I						I
Ptychoglyptus				W					W
Ptychopleurella				(W)				(W)	
Rafinesquina	Rh + IS	?			Rh + IS				
Rostricellula	C	C							C
Salopia		IS	IS	IS					
Skenidioides		W						W	W
Sowerbyella (S.)		W	W		W	W	W		W
Sowerbyites									(W)
Sulevorthis			I	I	I			?	
Tissintia	Rh	Rh							
Triplesia	C				C				C
Valcourea*			I						I

Regions and localities: An, Anglesey; Av, Avalonia (Shropshire, South Wales); Balt, Baltica (Estonia N, Oslo, central and south Sweden); HG, high-latitude Gondwana and peri-Gondwana (Algeria, Bohemia, Brittany, Montagne Noire, Morocco, Normandy, Portugal, Spain); MaS, South Mayo Terrane; Laur, Laurentian Platform (Canada E, Canada NW, USA C, USA E, USA NE marg., USA SE marg., USA W); Laur-deep, Laurentian Platform deep-water site (USA NE marg.-deep); LT, Leinster Terrane (Tramore and Dunabrattin); Prec-deep, Precordillera deep-water site (Precordillera-deep). *Scoto-Appalachian taxa.

Grorudia, Harknesella, Horderleyella, Leptaena, Platystrophia and *Salopia* (see Table 7). Most of the individual sites within the province contain a number of local endemic genera, which are described in the following sections. Several originations followed by migration are recorded from both East Avalonia and the Baltic Palaeobasin suggesting that well-developed species pumps were operating on both palaeoplates. The Anglo-Welsh–Baltic faunas are distinct from those of Laurentia and its margins except

for the deep-water faunas, which shared a significant number of genera with the province as described earlier.

The Leinster Terrane

The early Sandbian Leinster faunas were collected from the deeper-water argillaceous carbonate facies of the Tramore Limestone Formation and deep-water facies of the Dunabrattin Limestone Formation. All genera recorded from the Dunabrattin

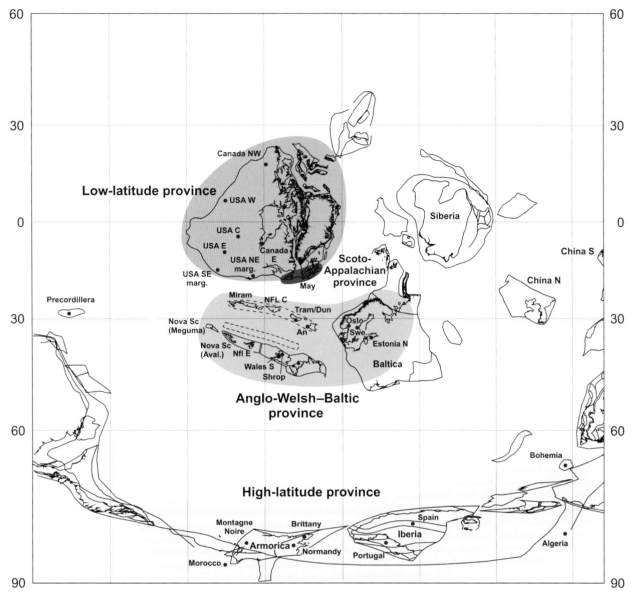

Fig. 14. Early Sandbian (457 Ma) provinces. The current study mainly focused on brachiopod provinciality within the Iapetus and Rheic oceans and the Tornquist Sea and, hence, the interpreted provinces are only shown for the investigated areas. The full extent of the provinces (excluding the Scoto-Appalachian Province) is given in Harper *et al.* (2013). Laurentian Platform localities: Canada E, east Canada; Canada NW, northwest Canada; USA C, central USA; USA E, east USA; USA NE marg., northeast margin of USA (incl. one shallow-water and one deep-water locality); USA SE marg., southeast margin of USA; USA W, west USA. Peri-Laurentian terrane localities: May, South Mayo Terrane. Precordilleran localities: Precordilleran platform. PVA-TEB terrane localities: An, Anglesey (Monian Composite Terrane); Tram/Dun, Tramore and Dunabrattin areas (Leinster Terrane). East Avalonian localities: Shrop, Shropshire; Wales S, South Wales. Baltic Platform localities: Estonia N, north Estonia; Swe, south and central Sweden; Oslo, Oslo region, Norway. Gondwanan and Peri-Gondwanan localities: Algeria, Bohemia, Brittany, Montagne Noire, Morocco, Normandy, Portugal, Spain. Other: China N, North China; China S, South China; Miram, Miramichi Terrane; NFL C, Central Newfoundland Terrane; NFL E, East Newfoundland terrane; Nova Sc, Nova Scotia.

Limestone are also present in the Tramore Limestone, although Dunabrattin diversity is lower. The early Sandbian Leinster fauna is of slightly higher diversity (18 genera) than it was during the Darriwilian (14 genera) and it only includes one possibly endemic genus, *Hibernobonites* n. gen., which was already present on the Leinster Terrane in the Darriwilian. The Leinster fauna maintained its mixed composition largely due to survival of the already present genera, and it did not receive new taxa from the High-latitude Province. Twelve of the reported genera were already present in the Darriwilian section of the Tramore Limestone and the six new genera include *Atelelasma*, *Glyptambonites*, *Grorudia*, *Isophragma*, *Leptestiina* and *Productorthis*. Of these, *Grorudia* has biogeographical value on generic and specific level and *Leptestiina* has biogeographical value on specific level. *Grorudia* (i.e. *G. grorudi*) is the only of the newly arrived genera that had its first occurrence in the early Sandbian and furthermore had a very restricted distribution limited to the Oslo Region and the Leinster Terrane. *Grorudia grorudi* likely originated within the Oslo Region of the Baltic Palaeobasin during the earliest Sandbian (Spjeldnæs 1957), from where it migrated to the southeastern Leinster Terrane (Tramore Lst. Units 3–5 and Dunabrattin Lst. Units III–IV). *Leptestiina derfelensis* co-occurs in Tramore and Anglesey (Bates 1968) in the early Sandbian and is recorded slightly later from the mid Sandbian in East Avalonia (Cocks & Rong 1989). This suggests that taxa now migrated from the Ganderian terranes into East Avalonia, which was not reported from the Darriwilian time-slice, but is in accordance with the effects of the *gracilis* drowning reported elsewhere. Another species, *Platystrophia* aff. *Platystrophia sublimis*, is of biogeographical importance, although the genus was already present in the Tramore Limestone during the Darriwilian. It may be a close descendant of *P. sublimis* from the early Sandbian of Estonia that evolved after migrating from the Baltic Palaeobasin to the Leinster Terrane. The remaining newly arrived genera are of earlier Ordovician Laurentian or possibly Baltic origin (but the record of *Glyptambonites* in Rasmussen 2005 requires taxonomical description and illustration), and they had a more or less widespread distribution prior to the *gracilis* drowning and may have migrated to the Leinster Terrane from a number of sites. It is likely, though, that at least some of the originally peri-Laurentian (i.e. Scoto-Appalachian) genera, recorded from the Leinster Terrane, migrated directly from the intraoceanic Laurentian sites, as the mixing of south and north Iapetan taxa was generally pronounced.

The Leinster fauna contains a significant number of typical Scoto-Appalachian taxa, that is *Atelelasma*, *Isophragma*, *Leptellina*, *Paurorthis*, *Productorthis* and *Valcourea*, of which the first three were recorded from the early Sandbian South Mayo Terrane, although *Atelelasma* was questionably assigned. *Leptellina*, *Paurorthis* and *Valcourea* were already present on the Leinster Terrane in the late Darriwilian. The appearance of the Scoto-Appalachian fauna in the deeper (deep DR 3–DR 4) carbonate facies of the Leinster Terrane is coincident with the *gracilis* drowning, which was also noted by Harper & Parkes (1989). The Leinster Terrane is, however, more closely related to the geographically nearby areas of Anglesey, East Avalonia and Baltica as indicated by the cluster and PCO analyses (Figs 12 and 13, respectively) and

the higher number of taxa (genera and species) shared between the Leinster Terrane and these sites. The Leinster Terrane shares seven non-cosmopolitan genera with the Anglo-Welsh–Baltic sites and a maximum of four with the South Mayo Terrane. Several of the Anglo-Welsh–Baltic taxa had a very restricted distribution within the southern Iapetus Ocean and Baltic Palaeobasin (i.e. *Grorudia*, *Platystrophia*, *Salopia*), while the Scoto-Appalachian taxa were widespread (see also Table 7) and not very good biogeographical indicators.

East Avalonia and Anglesey

The association between the East Avalonian faunas (Shropshire, South Wales) and the Ganderian Anglesey fauna is not unequivocally interpreted by the cluster and PCO analyses (Figs 12 and 13, respectively). The problem arises when the two analyses compare them with the deep-water faunas of the Laurentian Platform and peri-Laurentian margin, and while Anglesey is most closely related to the deep-water faunas in the cluster analysis, South Wales is most closely related to these in the PCO analysis. East Avalonia and Anglesey are generally closely related as well as to the rest of the Anglo-Welsh–Baltic Province sites. East Avalonia and Anglesey share up to three genera that are only recorded from the early Sandbian in East Avalonia and the PVA-TEB terranes, that is *Harknesella*, *Horderleyella* and *Salopia*, with *Horderleyella* being questionably identified in Anglesey. *Harknesella* may have originated in either Anglesey or Shropshire in the early Sandbian as it first occurs within the *gracilis* Zone of these sites (Bates 1968 and Williams 1974; respectively). Several other genera may have originated in East Avalonia in the early Sandbian including *Bystromena*, *Hedstroemina*, *Heterorthis* and *Salacorthis* (Williams 1974), of which *Bystromena* and *Salacorthis* have not been recorded outside East Avalonia and probably remained endemic to this area throughout their existence. *Hedstroemina* and *Heterorthis* migrated to the Oslo region and the High-latitude Province, respectively, later in the Sandbian (Spjeldnæs 1957; Havlíček 1970, 1977, respectively) indicating that East Avalonia operated as a species pump during the Sandbian.

East Avalonian diversity is high with a total of 25 genera and an endemicity of 36%, although local diversity is highly skewed with 21 genera recorded from Shropshire and only eight from South Wales. Lithologies from both localities include sandstones and shales and the distinct diversities are probably not related to substrate type. The total diversity and endemicity is very close to the mid–latest Darriwilian levels (22 genera, 32% endemicity), although now there is evidence for migrations out of Avalonia. Due to the low diversity of South Wales only few genera co-occur in the East Avalonian localities including *Dalmanella*, *Glyptorthis*, *Horderleyella* and *Sowerbyella* (*Sowerbyella*), of which *Dalmanella* apparently was endemic.

Anglesey comprises the last occurrence of endemic Celtic Province diagnostic taxa, that is *Jaanussonites*, *Monorthis* and *Rectotrophia* recorded from the upper *gracilis*–lower *foliaceous* pebbly grits of the Llanbabo Formation by Bates (1968). Even so, the Anglesey fauna is dominated by taxa widespread within and outside the Iapetus Ocean, as well as non-Celtic taxa endemic to the Anglo-Welsh–Baltic Province (see Table 7). It shares 12 genera with the Anglo-Welsh–Baltic localities. It does not contain any typical Scoto-Appalachian taxa even though several Scoto-Appalachian taxa are reported from the geographically close Leinster Terrane. Diversity of the early Sandbian Anglesey fauna is slightly higher than in the Darriwilian (24 and 18 genera, respectively) but endemicity remained about the same (21% and 17%, respectively) the latter largely owing to the presence of the Celtic diagnostic genera. Only one possible generic origination is recorded from the early Sandbian in Anglesey (i.e. *Harknesella*) but it is more likely that it originated in East Avalonia as several other originations are reported here. The trend reported for Anglesey is that the previous Celtic assemblages were replaced by more widespread taxa from both the north and south Iapetus-bordering areas as a result of the *gracilis* drowning.

Baltica
The early Sandbian Baltic Platform faunas are still somewhat distinct in the cluster and PCO analyses (Figs 12 and 13), although they form part of the larger cluster encompassing the transitional faunas of the Anglo-Welsh–Baltic Province. The Baltic Palaeobasin still contained several endemic taxa, however, some of which continued from the Darriwilian into the Sandbian and some of which originated on the platform during the early Sandbian.

The faunal distinctness between the Baltic confacies belts became more subtle with the *gracilis* drowning and several taxa commonly co-occurred in the Baltic sites including *Actinomena*, *Bilobia*, *Chonetoidea*, *Christiania*, *Cyrtonotella*, *Kullervo*, *Leptestia*, *Nicolella*, *Oepikina*, *Oxoplecia*, *Platystrophia*, *Septomena*, *Septorthis*, *Sowerbyella* (*Sowerbyella*) and *Tetraodontella*. Several genera originated on the Baltic Platform during the early Sandbian including *Grorudia* (Spjeldnæs 1957), *Gunnarella* (Hansen 2008), *Oslomena* (Spjeldnæs 1957), *Kurnamena* (Rõõmusoks 2004), *Septorthis* (Hansen 2008),

Tallinites (*Kukrusena* in Rõõmusoks 2004) and *Tetraodontella* (Jaanusson 1963; Cocks & Rong 1989; Hints & Rõõmusoks 1997). *Septorthis* may have originated on the platform during the late Darriwilian (Harper *et al.* 2004 unpublished database on first and last occurrences) although this is not confirmed by the present data. The majority of these genera remained endemic to the Baltic Platform throughout their existence. *Grorudia* migrated to the Leinster Terrane later in the early Sandbian as described above; *Gunnarella* has been recorded from the Sandbian–early Katian (Caradoc) of Tien Shan (Cocks 2005) and *Tetraodontella* has been recorded from South China possibly as early as the late Sandbian (Cocks & Rong 1989). Like East Avalonia, the Baltic Palaeobasin may have operated as a species pump during the Sandbian supplying new brachiopod taxa to other parts of the Late Ordovician World.

The *gracilis* drowning was responsible for migration of brachiopod taxa to the platform, and while this may have caused the major increase in diversity in the Norwegian (Oslo) and central and south Swedish sections (18 and 17 genera, respectively) compared with the early mid–latest Darriwilian interval (6 and 4 genera, respectively), diversity in North Estonia decreased slightly from 34 genera in the mid–latest Darriwilian to 28 in the early Sandbian. Endemicity in North Estonia and Oslo likewise decreased from 65% to 46% and 33% to 17%, respectively. Endemicity in the Swedish area was still extremely low (0.6%) and is likely not significant. The diminished North Estonian diversity is related to an increased number of disappearances from the area during the mid–late Darriwilian compared with localized originations and immigration into the area in the early Sandbian. Several endemic taxa (i.e. *Apatorthis*, *Hemipronites*, *Krattorthis*, *Lacunarites*, *Lycophoria*, *Panderites*, *Plectambonites* and *Tallinites*) and two non-endemic genera that also occurred in arc terranes (i.e. *Inversella* and *Noetlingia*) became extinct sometime before the end Darriwilian (Hints & Harper 2003), while others disappeared from North Estonia but persisted in other areas outside the Baltic platform. The Darriwilian extinctions together with changes in substrate type from limestone facies with less siliciclastic content to more siliciclastic-dominated facies made way for migration of taxa from other areas into the basin. The decrease in endemicity in North Estonia may be explained by the mid–late Darriwilian extinctions among endemic genera as well as immigration of more widespread taxa into the area in the early Sandbian. Decreased endemicity in the Oslo region was largely a function of the latter.

The High-latitude Province

The integrity and distinctive character of the high-latitude Gondwanan and peri-Gondwanan faunas are confirmed for the early Sandbian by the cluster and PCO analyses (Figs 12 and 13, respectively). The High-latitude Province likely operated as a species pump indicated by a high number of originations in the early Sandbian, primarily within the dalmanellidines, of which several are recorded from other parts of the World later in the Sandbian, for example *Drabovia* (e.g. Villas 1992; Botquelen & Mélou 2007), *Reuschella* (Villas 1992; Villas *et al.* 2011) and *Saukrodictya* (Havlíček 1977; Villas 1992). Several taxa commonly co-occur within the province including *Aegiromena*, *Drabovia*, *Gelidorthis*, *Heterorthina*, *Howellites*, *Jezercia*, *Rostricellula*, *Svobodaina*, *Tenuiseptorthis* and *Triplesia*, of which *Aegiromena*, *Heterorthina*, *Jezercia*, *Svobodaina* and *Tenuiseptorthis* were endemic.

The faunas of North Africa (Algeria, Morocco) and Amorica (Brittany, Montagne Noire, Normandy) are of low diversity including only 1–8 genera, while the Iberian faunas (Spain, Portugal) are of slightly higher diversity with 10–12 genera, and Bohemia is of highest diversity including 18 genera. The total diversity for the early Sandbian province is high and includes 25 different genera with a total endemicity of 56%, both of which are higher than the recorded Darriwilian levels (15 genera and 40% endemicity). The higher diversity was a function of migration of mainly globally widespread and cosmopolitan taxa into the province, facilitated by the *gracilis* drowning as well as the high number of originations, while the higher endemicity was a function of the latter. The significantly higher diversity in Bohemia was most likely not a function of facies changes favouring brachiopod settlement and preservation, as the sediments from which the taxa were recorded were still composed of quartzites, shales and ironstones as they were during the Darriwilian.

The Precordillera

The deep-water Precordilleran fauna is of very low diversity and only includes the two widespread genera *Dactylogonia* and *Sowerbyella* (*Sowerbyella*), which are also recorded from several other Iapetus-bordering sites (see Table 7). The low diversity may be related to poor preservation and/or ecological conditions not favouring brachiopods. The Las Aguaditas Formation consists of platy limestone with abundant insoluble residue, some breccias and slumped horizons (Astini 2003) suggesting that the

sediments may have been subject to diagenetic dissolution and tectonic activity.

Mid Sandbian deep-water siliciclastic formations from the San Juan Region include a total of 18 genera (Benedetto 2003b) showing that conditions favouring brachiopods were later re-established. These faunas contain several taxa of Gondwanan affinity, indicating a closer geographical position to West Gondwana than in the earlier part of the Darriwilian.

Conclusions

1 The two lowermost units of the Tramore Limestone Formation were correlated with the upper *murchisoni* and *teretiusculus* graptolite zones of late Darriwilian age, while the upper four units were correlated with the lower Sandbian *gracilis* Zone. The Dunabrattin Limestone Formation was likewise correlated with the *gracilis* Zone.

2 The first four units of the Tramore Limestone were deposited during continuously rising sea level correlating with the *gracilis* drowning, culminating in a highstand in Units 3–4 corresponding to the Furudal Highstand in Baltica. The two upper units were deposited during a global shallowing and lowstand correlated with the Vollen Lowstand in Baltica.

3 New Depth Range zones were established to better facilitate palaeobathymetric characterization of faunas. These were based on lithology, sedimentological structures and preservation of fossils relating to syndepositional mechanical processes and are more readily applied to any assemblage regardless of time, space, evolution and ecological adaptation among the brachiopods compared with the benthic associations (BAs) of Boucot (1975).

4 New reconstructions of the palaeoplate configuration in the mid Darriwilian and early Sandbian Iapetus Ocean place the Leinster Terrane on the leading edge of Ganderia together with the Monian Composite Terrane (including Rosslare and Anglesey), and the Bellewstown, Grangegeeth, Central Newfoundland and Miramichi terranes in association with the volcanic Popelogan-Victoria arc (PVA) several hundred kilometres north of Avalonia. This is in contrast to earlier interpretations positioning the Irish-English terranes on the leading edge of East Avalonia, and the North American-Canadian terranes either by an isolated volcanic arc in the Iapetus Ocean far north of Avalonia, or on the leading edge of

West Avalonia. The Tetagouche-Exploits backarc basin (TEB), which was situated between the leading and trailing Ganderia margins, progressively widened through the Ordovician and separated the Ganderian terranes from the trailing Ganderia margin and Avalonia.

5 Avalonia and Ganderia occupied a mid-latitude transitional position between the Iapetus and Rheic oceans and the Tornquist Sea in the early mid Darriwilian – early Sandbian, comprising settling grounds for high-, low- and mid-latitude genera. Avalonia and Ganderia were positioned closer to Baltica and Laurentia in the early mid Darriwilian – early Sandbian compared to the more distant high-latitude peri-Gondwana. This is indicated by the significantly higher proportion of taxa that migrated from Baltica and the Laurentian marginal and intraoceanic sites to the Ganderian terranes, and from the Laurentian marginal and intraoceanic sites to East Avalonia, compared to the low proportion of high-latitude genera reported from the Avalonian and Ganderian faunas. The Leinster Terrane furthermore shared some species with the Laurentian Platform, the peri-Laurentian Scottish Midland Valley Terrane, East Avalonia and Baltica in the late Darriwilian and the Laurentian Platform, East Avalonia and Baltica in the early Sandbian confirming its close relationship with the other Iapetus-bordering sites. Migrations within and between the investigated regions typically happened during pronounced drowning events and periods of global sea-level highstand.

6 The significant Laurentian component in the late Darriwilian and early Sandbian Tramore Limestone brachiopod faunas is not necessarily indicative of direct migration from the Laurentian region to the Leinster Terrane. The Darriwilian 'Laurentian' taxa (which did not necessarily originate in Laurentia but likely migrated from other sites via Laurentia and associated arc terranes) were either present in East Avalonia or on the other Ganderian terranes prior to deposition of the Tramore Limestone, and most of the early Sandbian (Scoto-Appalachian) genera were widespread prior to their occurrence on the Leinster Terrane. A few Laurentian/peri-Laurentian species have only been reported from the late Darriwilian Leinster Terrane outside the Laurentian region, however, suggesting that at least some taxa may have migrated directly during the Darriwilian. No Laurentian/peri-Laurentian species were recorded from the early Sandbian Leinster Terrane, although this does not preclude that

migration took place directly between these sites. The exchange of fauna between the northern and southern Iapetus-bordering regions was generally pronounced due to the *gracilis* drowning and their relatively close mutual geographical positions.

7 All Ganderian terranes, except the Leinster Terrane, comprised Celtic faunas in the Darriwilian and, hence, the Leinster Terrane was not included in the Celtic Province. Likewise, East Avalonia was not included in the Celtic Province as the faunas from Shropshire and South Wales did not contain any Celtic diagnostic genera, and the two sites shared very few taxa with any of Celtic localities. Localities belonging to the Celtic Province that were previously associated with East Avalonia include Anglesey and Rosslare and their present inferred position by the leading edge of Ganderia limits the latitudinal extent of the Celtic Province to 10–36°S.

8 The Central Newfoundland Terrane and Anglesey constituted a museum for the Celtic fauna as Celtic diagnostic taxa persisted in these sites into the early–mid Sandbian. The last occurrence of Celtic diagnostic taxa was recorded from Anglesey.

9 Oceanic surface currents were responsible for dispersal of the shelf-bound Darriwilian rhynchonelliform brachiopod larvae. During the early Darriwilian glaciation, temperatures and presumably atmospheric CO_2 levels were lower than in the remainder of the Darriwilian interval. This may have affected the flow direction of the oceanic currents resulting in dispersal of brachiopod taxa from, for example, Laurentia in the east to the Precordillera in the west. The following mid–late Darriwilian warming event and rise in atmospheric CO_2 levels changed oceanic circulation resulting in currents flowing from West Gondwana and eastwards across the Precordillera into the Iapetus Ocean. Furthermore, dispersal of brachiopod taxa from the Laurentian marginal and intraoceanic sites to the southern Iapetus–Tornquist region may only have happened indirectly in the early Darriwilian, as currents are predicted to have flown west of the Avalonian/Ganderian segment. In the mid–late Darriwilian, the higher atmospheric CO_2 levels, probably together with other unknown factors, changed this flow pattern resulting in currents flowing directly from the Laurentian margin and intraoceanic sites to Ganderia. The brachiopod migrational patterns reported in the present study indicate that currents likely flowed from the Baltic Palaeobasin towards Ganderia during the Darriwilian. This is not supported by other Darriwilian oceanic

circulation models and additional fossil data and modelling is needed for a more comprehensive evaluation of this hypothesis.

10 Brachiopod provinciality was pronounced in the early mid–latest Darriwilian owing to the more widely dispersed palaeoplates and lower global sea level compared with the early Sandbian. The *gracilis* drowning increased global sea level in the early Sandbian, dispersing deeper-water biofacies across the shelfal areas. This resulted in significant brachiopod migrations on a global scale causing a rise in diversity in most of the investigated sites except in areas containing a significant proportion of inferred highly specialized taxa that either became extinct or disappeared from the area but continued elsewhere.

11 The number of major species pumps in the early mid–latest Darriwilian was possibly limited to the Laurentian and Baltic platforms. Main species pumps of the early Sandbian include several areas such as the Laurentian and Baltic Platforms, East Avalonia and high-latitude Gondwana and peri-Gondwana.

Systematic palaeontology

The higher taxonomical ranks follow the *Treatise on Invertebrate Paleontology Part H* (vol. 2–4, 2000–2002) unless stated otherwise. They are, hence, not included in the reference list to avoid unnecessary duplication.

The terminology applied herein is that of Williams & Brunton (1997) in the *Treatise on Invertebrate Paleontology Part H* (vol. 1), with exceptions described below. The terms lateral, posterior and anterior profile refer to the exterior valve profiles viewed from the side, posterior and anterior of the shell, respectively.

There has been some confusion through time concerning the use of the terms notothyrial platform, septalium, sessile septalium and sessile cruralium. Zuykov & Harper (2007, pp. 15–18) discussed this in their revision of the genus *Platystrophia* and their definitions are followed here for all the described genera of this study:

1 A notothyrial platform is defined as an umbonal thickening of the dorsal valve floor between the inner socket ridges, brachiophores or crural plates.

2 A septalium is defined as a troughlike structure in the dorsal valve between the crural bases consisting of crural plates or homologues fused medially and usually supported by a median septum but may be unsupported. A septalium is Y-shaped in cross-section if supported by a median septum or U-shaped without connection to the valve floor if unsupported. It does not carry adductor muscles.

3 A sessile septalium is defined as the septalium except it is resting on the valve floor. The term 'sessile cruralium' has in some instances been used instead of 'sessile septalium' in the discussion of the genus *Platystrophia* but a cruralium *sensu stricto* carries adductor muscles (Williams & Brunton 1997, p. 428).

Abbreviations. – Abbreviations used in the synonymy lists are as follows: *pars* = partly; *sic* = quotation;   aff. = affinity;   cf. = confer;   non = the present authors disagree with the species and/or genus assignment; ? = uncertain assignment.

Repository. – The illustrated specimens are reposited in the collections of the Natural History Museum, London, UK (NHMUK).

Statistical methods

A number of the genera and species were collected from both the shallow-water Tramore and deep-water Dunabrattin Limestone formations. Carlisle (1979) described the brachiopods in the deep-water facies as small and immature and Sleeman & McConnell (1995, p. 13) used this as an indicator for muddy environments. It was generally not possible to distinguish the shallow- and deep-water specimens in the unmarked samples as most species included a wide range of growth stages and, hence, they were treated as belonging to one sample in the statistical analyses.

All statistical tests were performed with the statistical software package PAST (Hammer & Harper 2006; Hammer *et al.* 2014). Analyses involving muscle scars are based on the complete muscle scar (or muscle field) concerning ventral valves and the complete adductor scar (or adductor field) concerning dorsal valves unless stated otherwise. Measurements on the Tramore and Dunabrattin specimens are provided in Appendix B.

Allometric growth

Brachiopods of a given species may display very different morphologies depending on growth stage. It is important to know the complete morphological range of a species in order to describe it properly as well as for future identification of the species from other collections. A number of new species may wrongfully be erected if a collection only includes certain growth stages of a poorly known species subject to allometric growth. When enough material was available, the new species described herein were therefore tested for allometry. Allometric growth *sensu stricto* refers to a change in proportion through ontogeny that is well described by a mathematical model for differential growth (Hammer & Harper 2006) and results in shape changes, such as changes from transverse valves in immature specimens of a given brachiopod species to equidimensional valves in mature specimens, corresponding to a change in relative growth rate between the valve width and length.

The morphometric reduced major axis regression analysis (RMA) was applied for this purpose and performed on logarithmic values of the analysed variables. It is not uncommon for morphometric data to have a lognormal distribution, meaning that log transformation will bring the data into a normal distribution, allowing the use of parametric tests that assume normality (i.e. RMA) (Hammer & Harper 2006). The method tests the significance of the slope (a) of the regression line for data points plotted in an x–y-coordinate system. Isometric growth is indicated when $a = 1$ and $P(a = 1)$ is equal to or larger than the significance level (=0.05 in this study). $a = 2$ indicates positive allometry in the form of a parabolic relationship and $a = 0.5$ indicates negative allometry in the form of a square root relationship (Hammer & Harper 2006). The variables chosen for allometric analysis using the log-transformed RMA should be independent to produce correct results. If the variables are not independent, allometry may still be assessed visually in a non-transformed RMA plot for indications of shape changes with increased valve size, but the interpretations are not supported by statistical results.

If the relative growth rate of two variables such as valve width and length changes significantly at a certain point during ontogeny, the regression line is deflected by the resulting change in the allometric coefficient (the slope (a)), the point of deflection being termed a breakpoint. The change in slope may be discrete making it necessary to divide the data into size-groups and compare their respective slopes by bivariate statistical methods (F- and t-tests) to assess whether they are significantly different, and the inferred breakpoint is a true breakpoint. This should only be performed on linearly correlated data as the regression line is less significant for uncorrelated data.

An important point to note regarding the RMA method is that if the analysed variables switch place so the one previously analysed as the x-coordinate is now the y-coordinate and *vice versa*, the slope of the regression line may change significantly, which

changes the outcome of the RMA. This investigation encountered a few cases where an analysis indicated both allometry and isometry depending on which variable was plotted on a given axis. In these instances, the growth patterns were assessed by studying non-transformed RMA plots of the variables' corresponding percentage ratios (e.g. the valve length: width ratio, when analysing valve width against valve length) plotted against ontogenetic stage represented by the valve length in this study. Allometric growth was rejected when the data points were randomly and widely scattered around the regression line for all growth stages with no detectable changes in distributional pattern; that is, breakpoints were absent.

Not all the new species described herein had equally long ventral and dorsal valves, but the valve lengths and widths for both valves of a given species were pooled together to increase the amount of data for the allometric analyses under the assumption that corresponding valves in the living specimens must have grown at equal rates to fit together for proper protection.

Phylum Brachiopoda Duméril, 1806

Subphylum Rhynchonelliformea Williams, Carlson & Brunton, 1996

Class Strophomenata Öpik, 1934

Order Strophomenida Öpik, 1934

Superfamily Strophomenoidea King, 1846

Family Strophomenidae King, 1846

Subfamily Strophomeninae King, 1846

Genus *Tetraphalerella* Wang, 1949

Type species (by original designation). – *Tetraphalerella cooperi* Wang, 1949, from the upper Elgin Limestone, upper Katian (middle *TS*.5d–*TS*.6b), Iowa, USA.

Discussion. – The morphological terminology and taxonomical classification of Cocks & Rong (2000) is followed here and the Tramore specimen is assigned to *Tetraphalerella*? rather than *Strophomena* based on its ventral adductor scar which is entirely enclosed by the diductors. According to Cocks & Rong (2000), the dorsal valve of *Tetraphalerella* has socket plates which are posterolaterally curved unlike *Strophomena* which has anterolaterally directed socket plates. The posterolateral curvature of the socket plates can be quite variable in some taxa, however (C.M.Ø. Rasmussen personal communication 2011), and should be used with care in taxonomical classifications.

Tetraphalerella is currently suppressed as a subgenus under *Strophomena* (Cocks & Rong 2000). Wang (1949) originally proposed to separate *Tetraphalerella* from *Strophomena* based, among other features, on the thinner socket ridges and more prominent and well-arranged pseudopunctae of *Tetraphalerella*. Dewing (1999, 2004) suggested placing *Tetraphalerella* within its own family and he distinguished *Tetraphalerella* from *Strophomena* on the occurrence of radially arranged pseudopunctae lacking taleolae in the former versus more irregularly arranged pseudopunctae with a smooth taleolate core in the latter. The Tramore specimen is not well enough preserved to study pseudopunctae, but Rasmussen *et al.* (2012) raised *Tetraphalerella* to genus level, which is followed here as the interior features of *Tetraphalerella*? seem distinct from *Strophomena*.

Rasmussen *et al.* (2012) recorded characteristics of both *Tetraphalerella* and *Strophomena* in the late Darriwilian – early Sandbian specimens they assigned to *T. planobesa* (Cooper, 1956). Their materials have cardinalia conforming to Wang's definition of *Tetraphalerella* together with the uncored pseudopunctae described by Dewing (1999, 2004). The pseudopunctae in their specimens are irregularly arranged, however, as in *Strophomena*. Rasmussen *et al.* (2012) suggested their material might belong to an ancestral form of *Tetraphalerella* and *Strophomena* but their material was too fragmentary to form the basis for a new genus and the Tramore material is also insufficient for this purpose. Mitchell (1977) described and figured *Strophomena cancellata* (Portlock, 1843), from the Killey Bridge Formation, upper Katian, Pomeroy, North Ireland, which also shows characteristics of both *Tetraphalerella* (enclosed ventral adductor tract) and *Strophomena* (dorsal socket plates slightly recurved to run parallel with hingeline). Pseudopunctae were not described, but if they are irregularly arranged as in the material of Rasmussen *et al.* (2012), *T. planobesa* and *S. cancellata* may belong to a new genus or the definition of *Tetraphalerella* and *Strophomena* should be revised.

Tetraphalerella was previously considered a mid–late Katian genus endemic to Laurentia and it was recorded from this interval by several authors (Wang 1949; Howe 1988; Jin *et al.* 1997; Jin & Zhan 2000). However, Nikitin *et al.* (2003) described it from the early Katian of Kazakhstan and Rasmussen *et al.* (2012) from the upper Darriwilian – lower Sandbian of west-central Alaska. Candela (2003) described *Strophomena* cf. *S. medialis* Butts, 1942; from the upper Sandbian of Pomeroy, North Ireland, which,

judging from his figured specimens (pl. 1, figs 17, 18) appears to have the enclosed ventral adductor tract and recurved dorsal socket plates of *Tetraphalerella*. Rasmussen *et al.* (2012) suggested the oldest occurrence of *Tetraphalerella* may date back to the early Sandbian. If the generic assignment of the Tramore specimen to this genus is confirmed, this study takes the origin of *Tetraphalerella* as far back as the late Darriwilian (*TS.4c*), that is the age of the lower part of the Tramore Limestone Formation from which this specimen was collected.

Tetraphalerella? sp.

Plate 1, figure 12

Material. – One ventral valve interior.

Description. – The available ventral valve is subcircular, 27 mm long and 34 mm wide with maximum width at the hingeline, slightly alate on one side and cardinal angle slightly less than 90° on the other side; lateral margins almost straight converging on strongly rounded and almost pointy anterior margin; gently resupinate with a rounded geniculation increasing the concave attitude of the valve margin; 78% as long as wide with geniculation occurring at 87% of the valve length anterior to the umbo; interarea apsacline to almost anacline with well-developed arched apical pseudodeltidium showing conspicuous growth lines.

Ventral interior with short, stout teeth; dental plates widely divergent with an angle of 135°, continuous with a thick, elevated rim that form the lateral boundaries of the subpentagonal ventral muscle scar being 85% as long as wide and 40% as long as the valve; diductor scars large, flabellate and rhomboidal, entirely enclosing the adductor scars; adductor track situated medially, broadly triangular, about as wide as long and with a marked, rounded median callist; mantle canal system only impressed on the subperipheral rim giving it a beaded appearance.

Discussion. – Candela (2003) noted that the shape and proportions of the ventral muscle scar is constant within species of *Strophomena* but highly variable between the species of this genus. As *Strophomena* and *Tetraphalerella* are very similar except for the enclosed ventral adductor scar and strongly recurved dorsal socket plates of the latter, the shape and proportions of the ventral muscle scar may also be diagnostic in species belonging to *Tetraphalerella*. Candela (2003) morphometrically defined ventral muscle scar shapes from Cooper's (1956) figured specimens, and *Tetraphalerella*? sp. of this study clearly belongs to his Type 1 (diamond-shaped muscle scar).

The Tramore specimen is compared to Darriwilian – early Katian Iapetus species traditionally assigned to *Strophomena* but with the enclosed ventral adductor scar and recurved dorsal socket plates as in *Tetraphalerella sensu stricto* in addition to diamond-shaped ventral muscle scar. Of these, the species most similar to *Tetraphalerella*? sp. are *Strophomena aubernensis nasuta* Cooper, 1956; from the Tyrone Formation, middle–upper Sandbian (*TS*.5b–lower *TS*.5c), Kentucky, USA; *Strophomena* cf. *S. medialis* described by Candela (2003), from the Bardahessiagh Formation, upper Sandbian (lower *TS*.5b–lower *TS*.5c), Pomeroy, Co. Tyrone, North Ireland; and *Strophomena steinari* Spjeldnæs, 1957, from the Furuberget Formation (previously *Coelosphaeridium* beds), middle–upper Sandbian (*TS*.5b–lower *TS*.5c), Ringsaker District, Norway.

Tetraphalerella? sp. appears to be distinct from all the comparable species in having the most widely divergent dental plates with an angle of 135°, together with the shape of its ventral diductor scar, which is more widely splayed than observed in any of the other species and completely enclosing a triangular adductor track situated around a distinct median callist. In the other resembling species, the ventral adductor tract appears to be oval or elongately triangular with a less distinct or absent median callist. The Tramore specimen also has a distinct outline shape only observed in *Strophomena aubernensis nasuta*. It is comparable to *Strophomena* cf. *S. medialis*, *S. steinari* and *S. aubernensis nasuta* regarding length:width ratios, the onset of geniculation, and cardinal angles (could not be measured on *S. steinari*). Spjeldnæs (1957, p. 147) mentioned the ventral muscle impressions of *S. steinari* as being of *Tetraphalerella* type but also stated that his specimens show the typical *Strophomena* features up unto the very late stages of growth. Until more material is collected, it is not possible to assess whether the Tramore form belongs to a new species of *Tetraphalerella*.

Occurrence. – Tramore Limestone Formation, Unit 1 or 2, Barrel Strand area.

Subfamily Furcitellinae Williams, 1965

Genus *Dactylogonia* Ulrich & Cooper, 1942

Type species (by original designation). – *Dactylogonia geniculata* Ulrich & Cooper, 1942 from the Little Oak Formation, upper Darriwilian (lower–upper middle *TS*.4c), Cahaba Valley, Alabama.

Dactylogonia costellata n. sp.

Plate 1, figures 1–11; Plate 10, figure 2

Derivation of name. – Alluding to the finely costellate ornamentation.

Holotype. – NHMUK PI BB35216, interior mould of ventral valve, Tramore Limestone Formation (Pl. 1, figs 6, 7).

Material. – 11 ventral valves and 11 dorsal valves.

Diagnosis. – Subcircular to subquadrate *Dactylogonia* species with acute cardinal extremities, profile geniculately concavoconvex; ornament finely unequiparvicostellate with major costellae separated medially by 1–3 finer costellae.

Description. – Subcircular to subquadrate, geniculately concavoconvex *Dactylogonia* species; 8–17 mm long ($n = 22$) and 11–23 mm wide ($n = 21$) with maximum width along hingeline and lateral and anterior margins gently rounded; cardinal angles acute and slightly alate; ventral profile gently convex posterior to the pronounced, strongly rounded geniculation occurring at 72% the valve length anterior to the umbo (62–88%, $n = 8$); ventral valve 76% as long as wide (64–92%, $n = 10$) and 32% as deep as long (21–37%, $n = 5$); dorsal profile planar to slightly concave becoming strongly concave at the geniculation occurring at 85% the valve length anterior to the umbo (77–94%, $n = 8$); dorsal valve 73% as long as wide (63–86%, $n = 11$); ventral and dorsal interareas moderately wide, apsacline and anacline respectively with apical pseudodeltidium and chilidium observed on only one specimen; ornament unequiparvicostellate with counts of six and seven ribs per mm, respectively, on two dorsal exteriors at 5 mm anterior to the umbo; swollen costellae irregular and separated by 1–3 fine costellae medially but often become more regular anteriorly where they commonly are less conspicuous and alternate with fine costellae; radial ornament is cancelled by very fine concentric growth lines.

Ventral interior (Pl. 10, fig. 2) with short, stout teeth and poorly developed dental plates defining the posterolateral borders of the ventral, subpentagonal muscle scar; muscle scar deeply impressed, 96% as wide as long (83–115%, $n = 7$) and 38% as long as the valve length (34–45%, $n = 7$); diductors divergent, widest medially and narrowing to apices anteriorly with faint ridges developed along their lengths; adductor scars raised on a slight mound and not extending as far anteriorly as the diductors; tubercles developed posterolaterally and medially on one ventral interior; lemniscate mantle canal system impressed on one specimen.

Dorsal interior with cardinal process composed of two divergent lobes raised above a low notothyrial platform; the lobes are ankylosed anteriorly to a pair of widely divergent, curved socket ridges; extensions of these socket ridges converge medially to form the posterior end of a low, rounded myophragm extending forwards for over one-half the length of the valve; two pairs of low, rounded transmuscle septa occur, the posterior pair being short, parallel with the socket ridges and separating the posterior and the anterior adductor scars, and the anterior pair being longer and forming the inner boundaries of the anterior adductors; *vascular myaria* not impressed.

Morphological variation. – Dactylogonia costellata n. sp. was investigated for allometric growth. Valve length and width (*L*_valve and *W*_valve, respectively) included a sufficient amount of measurements for morphometric analysis.

The log-transformed RMA plot (Fig. 15) of the valve widths plotted against valve length indicate no breakpoints in the distribution of the data although there is a gap in valve lengths between the smaller and larger valves at 12.0–15.0 mm (~1.09–1.17 in the log-transformed RMA) separating the data points into two groups; a group including the smaller valves with valve lengths of 8.0–12.0 mm (*n* = 13) and one consisting of larger forms with valve lengths of 15.0–17.4 mm (*n* = 8) (squares and circles, respectively, Fig. 15). The log-transformed RMA shows that relative growth of length and width was isometric as $P(a = 1) > 0.05$ and the slope (*a*) is very close to 1. Hence, no statistically significant changes in growth rate between these variables took place through ontogeny and it was accordingly not necessary to compare the individual regression slopes of the two size-groups statistically. The gap in valve lengths indicates that the assemblage probably consists of two different populations of sorted valve sizes.

Discussion. – The Tramore form is assigned to *Dactylogonia* rather than *Oepikina* based on its relatively small, deeply impressed ventral muscle scar; *Oepikina* consistently has a very large, often weakly impressed ventral scar and the strong geniculation of the Tramore specimens is not generally associated with *Oepikina*.

Dactylogonia costellata n. sp. is morphologically distinct from all other described species of *Dactylogonia* and is only compared to the closely related species with emphasis on their most distinct differences.

Dactylogonia homostriata homostriata (Butts, 1942) was described by Candela (2003) from the Bardahessiagh Formation, middle Sandbian – lowermost Katian (lower *TS*.5b–lower *TS*.5c), Pomeroy, Co. Tyrone, North Ireland. It can be distinguished from the Tramore form by its more transverse outline with ventral valves being 56% as long as wide and the presence of a dorsal sulcus and ventral fold in the geniculate trail. *Dactylogonia* sp. described by Mitchell (1977) from the same formation has similar length:width ratio and also possess a dorsal sulcus and, hence, may be conspecific with *D. homostriata homostriata*.

Dactylogonia homostriata indiscissa (Williams, 1962) was described from the Balclatchie Mudstone Formation, upper Sandbian (lower *TS*.5b–lower *TS*.5c), Dow Hill, Girvan, Strathclyde, Scotland. *Dactylogonia costellata* n. sp. and *D. homostriata indiscissa* can be distinguished by their dorsal interiors, the interiors of the latter characterized by straight socket ridges and a shorter myophragm as well as the presence of a ventral median ridge, which is not observed in *D. costellata* n. sp.

Dactylogonia? *multicorrugata* (Reed, 1917) emend. Williams (1962), was described by Williams (1962), from the basal Ardwell Farm Mudstone Formation, upper Sandbian (lower *TS*.5b–lower *TS*.5c), Ardmillan Braes, Girvan, Strathclyde, Scotland. *D. costellata* n. sp. can be distinguished from this by its significantly more transverse outline, *D.*? *multicorrugata* being almost equidimensional with a length:width ratio of 94% measured on the dorsal valve figured by Williams (1962, pl. 19, fig. 20).

Dactylogonia obtusa Cooper, 1956, was described from the Lincolnshire and Arline formations, middle–upper Darriwilian (*TS*.4c), USA. It is the only one of Cooper's sixteen formally described species of *Dactylogonia* that has radial external ornament similar to that of *D. costellata* n. sp. They can be distinguished by their dorsal interiors *D. obtusa* having a nearly obsolete myophragm and stronger transmuscle septa, which occupy slightly different positions to that of *D. costellata* n. sp.

Dactylogonia richardsoni (Reed, 1917) was described by Harper (2006b, p. 146) from the Black Neuk Member of the Mill Formation, lowermost Ashgill (upper *TS*.5d–lower *TS*.6a), on the foreshore near Shalloch Mill, Girvan Strathclyde, Scotland. This species was assigned by (Harper 2006b, p. 146) to *Cyphomena* (*Cyphomena*) following Cocks (1978) but reassigned to *Dactylogonia* by Cocks (2008) without explanation. It is known from only the

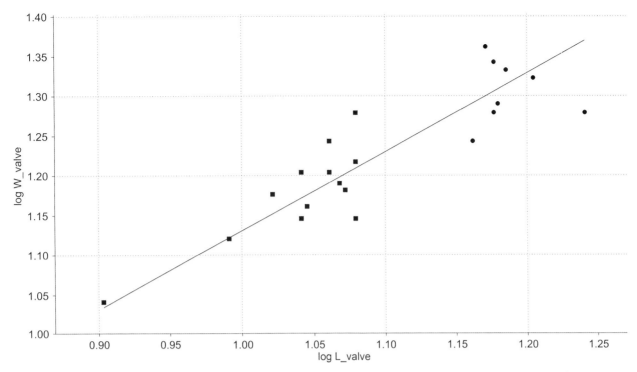

Fig. 15. Dactylogonia costellata n. sp. Log-transformed RMA of the valve width plotted against the valve length. $n = 21$, slope $a = 0.99$, error $a = 0.11$, $r = 0.88$, $P(\text{uncorr}) = 1 \times 10^{-7}$, $P(a = 1) = 0.93$, significance level = 0.05. Isometry cannot be rejected as $P(a = 1) > 0.05$ and the slope (a) is very close to 1.

lectotype and one other ventral valve. It can be distinguished from the Tramore form on its smaller size and finer ornamentation with 10 costellae per mm at 5 mm anteromedially to the umbo and the swollen costellae commonly being separated by four finer costellae as well as the presence of rugae.

Dactylogonia? *semiglobosina* (Davidson, 1883) was described by Williams (1962) from the Craighead Limestone Formation, middle Sandbian – lower Katian (lower *TS*.5b–upper middle *TS*.5c), Craighead, Strathclyde, Scotland. *D. semiglobosina* lacks a notothyrial platform and dorsal myophragm and it also has a less distinct and more rounded ventral muscle scar compared to *D. costellata* n. sp.

Dactylogonia sp. was described by Williams (1962) from the Confinis Flags and Auchensoul formations, uppermost Darriwilian (upper *TS*.4c), Girvan, Strathclyde, Scotland. It can be distinguished from the Tramore form by its earlier onset of dorsal geniculation occurring at 75% the valve length.

Bates (1968) described one poorly preserved and broken dorsal valve of *Dactylogonia* sp. from the Nantannog Formation, uppermost Darriwilian (upper *TS*.4c), Anglesey, North Wales. The Tramore and Welsh forms can be distinguished on their dorsal interiors the latter having more prominent transmuscle septa and an anteriorly bifurcating myophragm.

Occurrence. – Tramore Limestone Formation, Units 2–3, Barrel Strand area, Pickardstown section and Quillia Section 2. Dunabrattin Limestone Formation, Units I–III, Dunabrattin–Bunmahon area.

Family Rafinesquinidae Schuchert, 1893
Subfamily Rafinesquininae Schuchert, 1893

Genus *Colaptomena* Cooper, 1956

Type species (by original designation). – *Colaptomena leptostrophoidea* Cooper, 1956, from the Martinsburg Formation, uppermost Sandbian (lower *TS*.5c), Green Mount Church, Broadway Quadrangle, Virginia.

Discussion. – The distinction between the two very similar genera *Macrocoelia* Cooper, 1956 and *Colaptomena* Cooper, 1956 has resulted in some confusion. Recently Cocks & Rong (2000) synonymized these genera without explanation ranking *Colaptomena* as the senior synonym of the two.

Cooper (1956) provided a more detailed description of *Colaptomena* than *Macrocoelia* and in most cases he did not describe the same type of characters for both genera. Of the characters described for both, only the appearances of the ventral pseudodeltidium, the dorsal chilidium and the development of dorsal adductor ridges were observed to be different, although only slightly (Table 8). These characters can easily be obscured by weathering and also vary with ecology and they are not regarded as diagnostic here. Cooper (1956) further described *Colaptomena* as being smaller than *Macrocoelia* drawing the line at a valve length of 5 cm, but as valve size may vary significantly as a function of ecology, and considering that all the other described differences were unimportant, we do not regard size as being valid evidence for separating the two genera (Table 8). We therefore follow Cocks & Rong (2000) in suppressing *Macrocoelia* as a junior synonym of *Colaptomena*.

Cooper (1956) considered that the two genera succeeded each other stratigraphically separated by a small stratigraphical gap, *Macrocoelia* being the older occurring in the Darriwilian – early Sandbian and *Colaptomena* only occurring in the mid Sandbian–early Katian. We now consider the total range of *Colaptomena* to be at least Darriwilian – early Katian.

Colaptomena auduni n. sp.

Plate 1, figures 13–16; Plate 2, figures 1–8; Plate 10, figures 1, 4

Derivation of name. – In honour of Dr. Jan Audun Rasmussen, Mors, North Jutland, for his contributions to Ordovician research.

Holotype. – NHMUK PI BB35212, interior mould of ventral valve, Tramore Limestone Formation (Pl. 2, figs 1, 2).

Material. – Total of 19 ventral valves and eight dorsal valves.

Diagnosis. – Transverse to subquadrate, planoconvex to concavoconvex *Colaptomena* species with maximum width generally anterior to hingeline; maximum convexity of ventral valve located posteromedially, ventral valve 75% as long as wide; dorsal valve planar to gently concave, 74% as long as wide; ornamentation strongly unequiparvicostellate

and cancelled by fine concentric filae, irregularly swollen costellae separate 1–4 fine costellae; ventral muscle scar 86% as long as wide and dental plates diverging at an angle of about 104°.

Description. – Planoconvex to gently concavoconvex *Colaptomena* with transverse to subquadrate outline; 10–53 mm long (*n* = 27) and 15–71 mm wide (*n* = 24) with maximum width generally anterior of hingeline; cardinal angles commonly obtuse and lateral and anterior margins evenly rounded; ventral valve convex in lateral profile with maximum convexity posteromedially in adults and about medially in younger forms; 75% as long as wide (57–91%, *n* = 16) and 14% as deep as long (7–26%, *n* = 4); dorsal valve planar to gently concave, 74% as long as wide (64–84%, *n* = 8) and 15% as deep as long (12–17%, *n* = 2); ventral and dorsal interareas short, apsacline and anacline respectively; pseudodeltidium generally obsolete or occasionally developed apically, convex chilidium well-developed covering the cardinal process. External ornament unequiparvicostellate, and cancelled by dense, fine, concentric filae; irregularly swollen costellae separate 1–4 fine costellae and become less conspicuous anteriorly, counts of eight costellae per mm at 10 mm anteromedially of the umbo were consistent on six valves; impersistent rugae rarely developed but up to five pairs may occur posterolaterally at oblique angles to the hingeline.

Ventral interior (Pl. 10, fig. 4) with short, stout teeth; fine dental plates, widely divergent at an angle of about 104° (90–113°, *n* = 3), extending for a short distance anteriorly to form the posterolateral margins of the ventral diductor scars; small pedicle callist seen on some specimens with a short, fine median ridge extending anteriorly from it; ventral muscle scar large, subflabellate and weakly impressed with indistinct anterior margins, 87% as long as wide (71–102%, *n* = 8) and 43% as long as the valve length (37–52%, *n* = 8), diductor scars divided by six or seven faint, rounded ridges; up to three concentric low ridges may be developed between the convex median area and the flatter margins of the ventral valve interior; *vascula myaria* not impressed.

Dorsal interior (Pl. 10, fig. 1) with prominent, bilobed cardinal process; cardinal process lobes

Table 8. Key differences between *Colaptomena* and *Macrocoelia* (from Cooper 1956).

	Colaptomena	*Macrocoelia*
Size	Large; length and width almost 5 cm	Large; length and width more than 5 cm
Pseudodeltidium	Obsolete	Short
Chilidium	Narrow and sub-carinate	Large and convex
Dorsal adductor ridges	Absent	Subdued
Stratigraphic range	Mid Sandbian–early Katian	Darriwilian – early Sandbian

diverging at an acute angle from the umbo, subparallel-sided or slightly wedgelike in outline with flattened upper surfaces rise high above the notothyrial platform but become relatively thicker and lower in adult specimens; 78% as long as their combined width (74–91%, *n* = 7) and 9% as long as the valve length (7–11%, *n* = 7); low socket ridges ankylosed to the anchor-shaped notothyrial platform; dorsal adductors weakly impressed but two pairs of fine, faint transmuscle septa may be observed, the anterior pair extending beyond the myophragm; up to two low, subperipheral rims may be developed; *vascula myaria* indistinct but faint impressions of vascula genitalia were observed posterolaterally on one specimen.

Morphological variation. – The location of the maximum convexity in the ventral valves changed through ontogeny and was located posteromedially in the larger adults and about medially in smaller, younger forms. This may indicate that other morphological variables changed through ontogeny as well and the assemblage was investigated for allometric growth. The valve length and width (*L*_valve and *W*_valve, respectively) included a sufficient amount of measurements for statistical analysis.

The log-transformed RMA plot (Fig. 16) of the valve widths plotted against valve length indicates no breakpoints in the distribution of the data and no gaps in valve lengths. The assemblage was accordingly analysed statistically as one sample. The log-transformed RMA shows that length and width growth was isometric as $P(a = 1) > 0.05$ and the slope (*a*) is close to 1.

Discussion. – Two species of *Colaptomena* were recorded from the Tramore Limestone Formation in this study; *C. auduni* n. sp. and *Colaptomena pseudopecten*? M'Coy, 1846, although the latter is only represented by a single dorsal valve.

Of all the currently assigned species, *C. auduni* n. sp. is most similar to *C. pseudopecten* originally described by M'Coy (1846) from beds of probably early Sandbian age, Tramore area, southeast Ireland, and Grangegeeth, east Ireland (see discussion of *C. pseudopecten*? below). The ventral valves of *C. pseudopecten* illustrated by M'Coy (1846, pl. 3, fig. 22: length 25.0 mm, width 46.5 mm) and Cocks (2008, pl. 2, fig. 7: broken specimen, min. length 23 mm, min. width 33 mm) falls within the size range of the *C. auduni* n. sp. valves. They are comparable in their rectangular to subquadrate outline and their ribbing although the description given by M'Coy was not very precise. *Colaptomena pseudopecten* differs from

C. auduni in being more transverse with a valve length:width ratio of 54% and the maximum valve width occurring at the hingeline. The length:width ratio of the ventral muscle scar is likewise different between the two species being 60–65% as long as wide measured on the specimens illustrated by Cocks (2008) and M'Coy (1846). The angle between the dental plates seems to be generally larger in *C. pseudopecten* than in *C. auduni* n. sp. being about 110° in M'Coy's specimen and 130° in Cocks' specimen. No dorsal valves were described or illustrated by M'Coy (1846) or Cocks (2008) for *C. pseudopecten*. If the complete dorsal valve of *C. pseudopecten*? is conspecific with *C. pseudopecten*, the dorsal interiors are different compared to *C. auduni* n. sp. as the former has less massive cardinal process lobes and more pronounced dorsal socket ridges and furthermore lacks the subperipheral rims of *C. auduni* n. sp. The development of the myophragm and the two pairs of lateral septa are comparable.

The other comparable species of similar age are British and differ from *C. auduni* n. sp. in their costellate ornamentation and a few other individual morphological characteristics. The ornamentation of *C. llandeiloensis* (Davidson, 1871) (MacGregor 1961; latest Darriwilian, Wales) differs in not being markedly differentiated into primary and secondary costellae. The ornamentation of *C. concentrica* (Portlock, 1843) (Mitchell 1977; late Sandbian, North Ireland), *C. expansa* (Sowerby, 1839) (Williams 1963; Sandbian–early Katian, Wales), *C. macallumi* (Reed, 1917) (Williams 1962; Darriwilian, Scotland) and *C. prolata* (Williams, 1963) (late Sandbian, Wales) differs in being more coarsely costellate. *C. concentrica* has 5–6 costellae per mm at 10 mm anteromedially of the umbo, *C. expansa* has 4–5, *C. macallumi* has six and *C. prolata* has 6–7 costellae.

Occurrence. – Tramore Limestone Formation: Unit 2, Barrel Strand area, Pickardstown section and Quillia Section 1.

Colaptomena pseudopecten? (M'Coy, 1846)

Plate 2, figures 14, 15

?1846   *Orthis pseudopecten* M'Coy, p. 33, pl. 3, fig. 16.

?1978   *Macrocoelia pseudopecten* (M'Coy); Cocks, p. 114.

?2008   *Colaptomena pseudopecten* (M'Coy); Cocks, pl. 2, fig. 7.

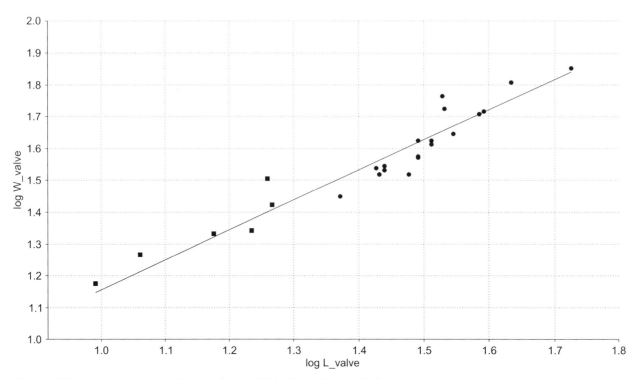

Fig. 16. Colaptomena auduni n. sp. Log-transformed RMA of the valve width plotted against the valve length. $n = 24$, slope $a = 0.94$, error $a = 0.06$, $r = 0.96$, $P(\text{uncorr}) = 3 \times 10^{-13}$, $P(a = 1) = 0.35$, significance level = 0.05. Isometry cannot be rejected as $P(a = 1) > 0.05$ and the slope (a) is very close to 1.

Material. – One dorsal valve.

Description. – The available dorsal valve is elongately semi-oval with evenly rounded lateral and anterior margins and acute cardinal angles; dorsal profile planar to slightly concave; 31 mm long and 62 mm wide with maximum width at straight hingeline; valve half as long as wide; interarea narrow and anacline with well-developed convex chilidium; ornamentation unknown.

Dorsal interior with prominent bilobed cardinal process; cardinal process lobes diverging at an acute angle from the umbo, subparallel-sided and slightly wedgelike in outline with flattened upper surfaces rise high above the notothyrial platform, 73% as long as their combined width and 6% as long as the valve length; low socket ridges ankylosed to the anchor-shaped notothyrial platform continuing medially into a low, rounded myophragm, 28% as long as the valve length measured from the umbo; a short thin median septum extends anteriorly from this and terminates at 45% the valve length anterior to the umbo; two pairs of weakly developed transmuscle septa runs subparallel with the myophragm, the posterior pair separating the posterior and anterior adductors and the anterior pair being almost twice as long as the myophragm and defining the inner boundaries of the anterior adductors;

adductor scars not impressed; subperipheral rims not observed.

Discussion. – *Colaptomena pseudopecten* was originally described and figured by M'Coy (1846) from beds of probably early Sandbian age (Caradoc *in* Cocks 1978, 2008), Tramore area, Co. Waterford, southeast Ireland, and Grangegeeth, Co. Meath, east Ireland. A lectotype from the Caradoc of the Tramore area was selected by Cocks (1978) and re-illustrated by Cocks (2008). Both M'Coy and Cocks only included ventral valves and, hence, dorsal valves are unknown. The valve shape of the *C. pseudopecten* specimen illustrated by M'Coy is very characteristic and similar to the dorsal valves described by this study with a valve size of 25.0 mm in length and 46.5 mm in width, a length:width ratio of 54% and by being widest at the hingeline with acute cardinal angles. The only difference in overall valve shape is the rectangular shape of the ventral valve of *C. pseudopecten* compared to the rounded dorsal valves from this study. No other species of *Colaptomena* or the very similar genus *Rafinesquina* has been recorded with valves this transverse. More material is needed to verify if the current form from the Tramore Limestone Formation is conspecific with *C. pseudopecten*.

Occurrence. – Tramore Limestone Formation, Unit 2, Barrel Strand area.

Superfamily Plectambonitoidea Jones, 1928
Family Plectambonitidae Jones, 1928
Subfamily Taphrodontinae Cooper, 1956

Genus *Isophragma* Cooper, 1956

Type species (by original designation). – *Isophragma ricevillense* Cooper, 1956; from the Athens Formation, lower Sandbian, Riceville, Tennessee, USA.

Isophragma parallelum n. sp.

Plate 2, figures 9–13; Plate 3, figures 1–4; Plate 10, figures 5–6

Derivation of name. – Alluding to the two pronounced, subparallel submedian septa.

Holotype. – NHMUK PI BB35251, interior mould of dorsal valve, Tramore or Dunabrattin Limestone Formation (Pl. 2, figs 11, 12).

Material. – Total of six ventral valves and five dorsal valves.

Diagnosis. – Elongately subcircular to semi-quadrate, weakly biconvex to planoconvex *Isophragma* species with acute to alate cardinal angles; ventral valve 46% as long as wide with sharp fold at umbo becoming weaker anteriorly; ornament of fine, branching costellae, sporadically swollen, seven per mm at 2 mm anterior to the umbo with branching occurring about 2 mm anterior to the umbo; ventral muscle scar 74% as long as wide extending anteriorly for 26% of the valve length; dorsal interior with two strong, subparallel submedian septa rising to a crest at the anterior margin of the muscle scar and extending for 82% the length of the valve.

Description. – Elongately subcircular to semi-quadrate, weakly biconvex to planoconvex *Isophragma* species; 3–6 mm long ($n = 11$) and 6–14 mm wide ($n = 11$) with maximum width along the hingeline; cardinal angles acute to alate and lateral and anterior margins broadly curved; ventral valve 46% as long as wide (39–52%, $n = 6$) with sharp fold at umbo becoming wider and weaker anteriorly; posterolateral areas flat; dorsal valve 42% as long as wide (31–49%, $n = 5$), strongly sulcate just posterior of the umbo with well-developed rounded flanks flattening posterolaterally; sulcus

weakening anteriorly becoming just distinguishable marginally in larger valves (valve length >4 mm); ventral interarea apsacline, with a short, narrow pseudodeltidium; dorsal interarea hypercline; ornament of fine branching costellae with swollen costellae occurring sporadically; seven costellae per mm at 2 mm anterior to the umbo (7–8, $n = 8$) and an average of 6.5 costellae per mm with a mode of 6–7 at the anterior margin for two dorsal and one ventral valve; interspaces about equal in width to the costellae; main branching from primary costellae occurs about 2 mm anterior to the umbo; strong concentric growth lines sporadically developed.

Ventral interior (Pl. 10, fig. 5) with short teeth and short dental plates, the latter enclosing the lateral margins of the muscle scar; muscle scar subtriangular, weakly impressed and undifferentiated, 73% as long as wide (67–83%, $n = 5$) and 26% as long as the valve length (24–30%, $n = 5$); rounded, w-shaped, raised area occurring about medially in the valve is 73% as long as wide (62–85%, $n = 3$) and 52% as wide as the maximum valve width (50–55%, $n = 3$) measured from the posterior margin; the raised area is cut by thin callosities developed anteriorly and laterally to the muscle scar and by straight, subparallel branches of the *vascula media*, the main pair of canals diverging from the ends of the diductor scars.

Dorsal interior (Pl. 10, fig. 4) with simple cardinal process on a short notothyrial platform; brachiophores short, ridgelike and widely divergent, 20% as long as their lateral spread (17–25%, $n = 4$) with lateral spread 22% of the valve width (21–23%, $n = 4$); dorsal adductor scar subrounded, 77% as long as wide (68–86%, $n = 4$) and 47% as long as the valve length (43–52%, $n = 4$), well impressed on a slightly raised bema defined anteriorly and laterally by a steep ridge; bema divided by a pair of strong, subparallel submedian septa arising at the base of the notothyrial platform and diverging very slightly to the anterior of the muscle scar where they rise to a crest and then continue anteriorly for 82% the valve length (78–86%, $n = 4$); the septa are not noticeably thickened in even the largest specimens although the area between them may be raised by deposits of secondary shell; on each side of the septa, the adductor scar is divided by several divergent ridges which become stronger towards the margins of the scar although not indicating a segregation into anterior and posterior parts; straight but faint mantle canals are observed from the lateral margins of the adductor scars.

Morphological variation. – *Isophragma parallelum* n. sp. was investigated for allometric growth. The valve length and width (*L_valve* and *W_valve*, respectively) included the largest amount of measurements ($n = 11$) of all the variables, although 11 specimens is rarely enough material to make representative statistical analysis.

The log-transformed RMA plot (Fig. 17) of the valve widths plotted against valve length shows that there is a gap in valve lengths between 4.1 and 5.3 mm (~0.62–0.73 in the log-transformed RMA plot) separating the data into two groups: one including smaller specimens with valve lengths of 2.5–4.1 mm ($n = 9$) and one consisting of large, adult forms with valve lengths of 5.3–5.7 mm ($n = 2$) (squares and circles, respectively, Fig. 17). Due to the small number of specimens in the assemblage it was not possible to detect if a breakpoint was present and, hence, if any sudden changes in growth rate took place. The small number of specimens makes the data unfit for statistical comparison of the individual regression slopes for the two size-groups and the total assemblage was analysed as one sample. The result of the log-transformed RMA indicates that length and width growth was isometric as $P(a = 1) > 0.05$ and the slope (a) is close to 1; however, more specimens are necessary to verify this.

Discussion. – *Isophragma* is rare in the Tramore and Dunabrattin limestones but the samples collected are sufficient to compare this species with others recorded elsewhere. None of the currently known species of *Isophragma* are remotely similar to *I. parallelum* n. sp. which in most cases can be distinguished by its extremely transverse valves, its submedian septa being very close to parallel and extending for the major part of the valve length, and its dorsal sulcus which fades out anteriorly.

The species most similar to *I. parallelum* n. sp. is *I. extensum tricostatum* Williams, 1962; from the Stinchar Limestone, uppermost Darriwilian (upper TS.4c), Girvan, Scotland. It is equal to *I. parallelum* n. sp. in size and has a comparable rib count of six costellae per mm anteromedially. *I. extensum tricostatum* is less transverse than *I. parallelum* n. sp. with ventral valves 58% as long as wide compared to 46% of *I. parallelum* n. sp. with a range of 39–52%. The submedian septa of *I. extensum tricostatum* are slightly divergent and shorter than in *I. parallelum* n. sp. their full extension being less than 75% the valve length. Like *I. extensum* Cooper, 1956, *s.s.* the subspecies has a dorsal sulcus passing anteriorly into a fold, the dorsal sulcus of *I. parallelum* n. sp. dying out anteriorly. In addition, *I. extensum tricostatum* derives its name from the thickening of the median and two sublateral costellae, which does not occur in *I. parallelum* n. sp.

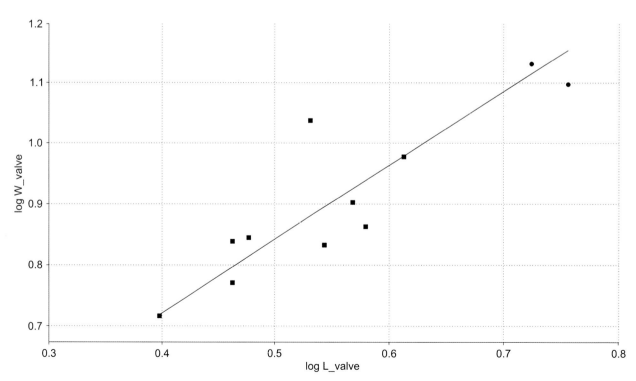

Fig. 17. Isophragma parallelum n. sp. Log-transformed RMA of the valve width plotted against the valve length. $N = 11$, slope $a = 1.21$, error $a = 0.19$, $r = -0.88$, $P(\text{uncorr}) = 3 \times 10^{-4}$, $P(a = 1) = 0.29$, significance level = 0.05. Isometry cannot be rejected as $P(a = 1) > 0.05$ and the slope (a) is close to 1.

Occurrence. – Tramore Limestone Formation, Unit V, Barrel Strand area. Dunabrattin Limestone Formation, Units III–IV, Dunabrattin–Bunmahon area.

Family Leptellinidae Ulrich & Cooper, 1936
Subfamily Leptellininae Ulrich & Cooper, 1936

Genus *Leptellina* Ulrich & Cooper, 1936

Type species (by original designation). – *Leptellina tennesseensis* Ulrich & Cooper, 1936, from the Arline Formation, uppermost Darriwilian – Sandbian, Friendsville, Tennessee, USA.

Leptellina llandeiloensis (Davidson, 1883)

Plate 3, figures 6–14

1843 *Leptaena duplicata* (Murchison); Portlock, p. 454.

1883 *Leptaena llandeiloensis* Davidson, p. 171 *pars*, pl. 12, figs 26, 26a, b, *non* figs 27–29.

1917 *Plectambonites llandeiloensis* (Davidson); Reed, p. 876, pl. 13, figs 32–34, pl. 14, figs 1–3.

1928 *Leptelloidea llandeiloensis* (Davidson); Jones, p. 477.

cf. 1977 *Leptellina* cf. *llandeiloensis* (Davidson); Mitchell, p. 72, pl. 13, figs 14–17.

1962 *Leptellina llandeiloensis* (Davidson); Williams, p. 164, pl. 15, figs 27–29, 37.

1978 *Leptellina* cf. *llandeiloensis* (Davidson); Cocks, p. 93.

cf. 1994 *Leptellina* cf. *llandeiloensis* (Davidson); Parkes, p. 156, pl. 11, figs 8–14.

2003 *Leptellina llandeiloensis* (Davidson); Candela, p. 34, pl. 5, figs 7–12.

2008 *Leptellina* (*Leptellina*) *llandeiloensis* (Davidson); Cocks, p. 78, pl. 4, fig. 1.

2010 *Leptellina llandeiloensis* (Davidson); Candela & Harper, p. 262, figs 6 m, p.

Comments on the synonymy list. – The specimens illustrated by Davidson (1883, pl. 12, figs 27–29) do probably not belong to *L. llandeiloensis*. The dorsal valves (figs 27, 29) possess a bema, which is more concave and narrow (56% as long as wide measured on fig. 29) compared to the Tramore form (43% as long as wide, range 31–54%) and the Girvan form (48% as long as wide measured from Williams' figured specimen (1962; pl. XV, fig. 27). The exterior illustrated by Davidson (1883, pl. 12, fig. 28) appears to be rami- or fascicostellate.

Material. – 30 ventral valves and 35 dorsal valves.

Description. – *Leptellina* species with outline shape changing through ontogeny from very transverse in young specimens to subcircular in mature individuals; 3–11 mm long ($n = 51$) and 6–20 mm wide ($n = 51$) with maximum width along straight hingeline; acute to obtuse cardinal angles but commonly 70–80°; lateral and anterior margins rounded; lateral profile concavoconvex with strongly convex ventral valve with maximum convexity occurring centrally, 65% as long as wide (46–84%, $n = 29$) and 47% as deep as long (30–62%, $n = 24$); dorsal valve evenly concave, 55% as long as wide (39–73%, $n = 20$) and 28% as deep as long (20–33%, $n = 5$); ventral interarea long, anacline to slightly apsacline, arched pseudodeltidium developed anteriorly covering delthyrium, no pedicle foramen; dorsal interarea short, hypercline, arched chilidium partially covers the notothyrium; ornament finely unequiparvicostellate but commonly obscured except for the accentuated costae which are well developed on the ventral valves, numbering 10 at 5 mm and 17 at 10 mm anterior to the umbo on one specimen; on the dorsal valve, 5–7 accentuated costae appear to originate at the umbo, of which the median costa is the strongest; on two dorsal exteriors fine costellae are preserved with 11 and 12 costellae per mm at 5 mm anterior to the umbo; thickened concentric lamellae are occasionally developed, strongest peripherally.

Ventral interior with subtriangular muscle scar of variable shape, 73% as long as wide (34–112%, $n = 24$) and extending anteriorly of the umbo for 27% of the valve length (15–36%, $n = 23$); adductor scars do not extend as far anteriorly as the diductor scars; coarse pustules are developed on a small diamond-shaped platform anterior of and between the muscle scar lobes; finer pustules occur around the anterior margin; teeth short, massive; thickened subperipheral rim sporadically developed; mantle canal system faint but appears to be lemniscate or saccate.

Dorsal interior with cardinalia composed of a relatively large, trifid cardinal process ankylosed anteriorly to the valve floor, and widely divergent socket ridges, subparallel to hingeline, which subtends deep, rounded sockets posteriorly; cardinalia 25% as long as wide (17–38%, $n = 23$) and 14% as long as the dorsal valve length (11–20%, *n* = 10); low notothyrial platform narrows to median septum which thickens anteriorly and bisects the bema; bema bilobed, striated, undercut and highly raised, highest medially at its junction with the median septum, 43% as long as wide (31–54%, *n* = 24) and 67% as long as the valve length (60–75%, *n* = 11); dorsal vascular system weakly impressed.

Discussion. – The Tramore specimens are regarded as conspecific with *L. llandeiloensis* (Davidson, 1883). Williams (1962) described *L. llandeiloensis* from the Balclatchie mudstones, middle Sandbian, Girvan, Scotland. He analysed the species statistically and compared it to other species of the genus. The only difference between the Tramore and Girvan specimens is a greater ventral valve depth (i.e. greater convexity) of the Tramore specimens being 47% as deep as the valve length (30–62%, *n* = 24) compared to 35% (*n* = 13) in the Girvan form. Some differences would be expected between the Scottish and southeast Irish specimens as the latter belonged to a slightly older stock and they were separated by the Iapetus Ocean. So far the two assemblages are considered conspecific as greater ventral valve convexity is not enough to merit subspecific distinction on its own. Candela & Harper (2010) described *L. llandeiloensis* from the Kirkcolm Formation, lower Katian (lower middle–uppermost *TS*.5c), Kilbucho, Southern Uplands, Scotland. It seems that this species is very long-ranging but the calculated means of their five specimens all lie close to those to the Tramore specimens (ventral valve: 74% as long as wide, 40% as deep as long; dorsal valve: 51% as long as wide, 25% as deep as long; ventral interior: muscle scar 70% as long as wide, 28% as long as the valve length). They note (p. 262) 'these specimens resemble *L. llandeiloensis* (Davidson, 1883) from Girvan (Williams 1962) and Pomeroy (Mitchell 1977; Candela 2003) in their lesser concavity than the other Girvan species described by Williams (1962). The ventral muscle scar length:valve length ratio is also similar to that of *L. llandeiloensis* from Girvan'.

The other two species of *Leptellina* known from Great Britain were described from the Girvan area by Williams (1962). *Leptellina rhacta* Williams, 1962; from the top of the Stinchar Limestone, lowermost Sandbian (*TS*.5a) is very distinctive in possessing a strong concentric ornament as well as in shape and outline. *Leptellina semilunata* Williams, 1962, from the top of the *Confinis* Flags, uppermost Darriwilian (upper *TS*.4c) and the top of the Stinchar Limestone can also be distinguished by its ornament with five strongly accentuated costae in the umbonal region.

This species also differs in the shape of the bema which is about 60% as long as wide and about 50% as long as the dorsal valve length.

Occurrence. – Tramore Limestone Formation, Units 2–3, Barrel Strand area. Dunabrattin Limestone Formation, Units I–III, Dunabrattin–Bunmahon area.

Subfamily Palaeostrophomeninae Cocks & Rong, 1989

Genus *Glyptambonites* Cooper, 1956

Type species (by original designation). – *Glyptambonites musculosus* Cooper, 1956, from the Oranda Formation, lower Sandbian, Linville Station, Virginia, USA.

Glyptambonites sp.

Plate 3, figure 5

Material. – Total of one, poorly preserved, ventral valve.

Description. – The material at hand consist of a subtriangular, gently convex ventral valve with a slight increase in convexity anteriorly; cardinal angles about 90°; lateral margins gently curved and anterolateral margins more strongly curved; length 27 mm and width 35 mm (incomplete).

Ventral interior with short teeth; narrow, elongate, subpentagonal ventral muscle scar with greatly reduced or obsolete dental plates and well-developed muscle bounding ridges, muscle scar 160% as long as wide and 46% as long as the valve length; anterior margin of the muscle scar weakly defined; diductor scars divergent, long but with anterior margins indistinct where they join the *vascula media*; subtriangular adductor scar confined to the delthyrial cavity but poorly impressed, elevated between the diductor scars on a small, low platform; a pair of large, semi-elliptical callosities developed medially in the valve just anteriorly of the end of the diductor scars, the main part of each being 71% as wide as long; these are cut, especially posteriorly, by the deeply impressed saccate mantle canal system; lateral branching from the median trunks of the *vascula media* first occurring at 29% the valve length anterior to the umbo where the main trunks converge, cutting deeply through the internal posterior areas of the two callosities; second branching occurs at about 46% the valve length anterior to the umbo where the secondary canal passes over the median area of the callosity branching again at its anterolateral boundary; further branching from the principal trunks of the vascula media occurs about 55% and 67% the valve length anterior to the umbo, all the secondary canals finally splitting into numerous small channels anteriorly; gonadal sacs weakly impressed lateral to the diductor scars and adjoining the hingeline.

Discussion. – This specimen closely resembles the genus *Glyptambonites* in its large size and narrow but elongate, subpentagonal muscle scar with well-developed muscle bounding ridges. The presence of large median callosities at the anterior end of the diductor scars is, however, a more common feature of *Titanambonites*, which is equal in size to *Glyptambonites* but has a much wider, flabellate ventral muscle scar. The presence of callosities was not mentioned in the definition of the genus *Glyptambonites* by Cooper (1956, p. 712) but they are known to sometimes be present in *Palaeostrophomena* a very similar but smaller genus (Cooper 1956, p. 701) indicating their occurrence may not be diagnostic. Furthermore, the space between the ventral diductor scars and the area anterior to the adductor scar appears to be occupied by a granulose area as is diagnostic of *Glyptambonites* according to Cooper (1956) although this is rather obscured by weathering.

Glyptambonites is a rare genus and often poorly preserved making comparisons difficult. *Glyptambonites* sp. was compared to all the well-established Darriwilian – Sandbian species of *Glyptambonites* but none of these appeared to be similar to this and they all lack the large median callosities observed in the Tramore form. Two species of *Glyptambonites* have been described from the British Isles, *G.* aff. *G. glyptus* Cooper, 1956; and *G. minor* Candela, 2003; and two from North America, *G. glyptus* Cooper, 1956; and *G. musculosus* Cooper, 1956. As there are not many species to compare with, only brief comparisons with all of them are provided below.

Glyptambonites glyptus was described by Cooper (1956) from the Effna, Rich Valley and Edinburg formations, upper Sandbian, Virginia, USA. It is very distinct and differs from the Tramore form in having a semi-elliptical outline and a rather short, subtriangular, weakly impressed ventral muscle scar with weakly developed muscle bounding ridges. *Glyptambonites* aff. *G. glyptus* was briefly described by Williams (1962) from the lower Stinchar Limestone Formation, uppermost Darriwilian (upper

*TS.*4c), Girvan, Strathclyde, Scotland. Williams only described and figured the dorsal ornamentation and dorsal interiors, and hence, it cannot be compared to the Tramore form.

Glyptambonites minor was described by Candela (2003) from the Bardahessiagh Formation, Member III, upper Sandbian (*TS.*5b–lower *TS.*5c), Pomeroy, Co. Tyrone, Northern Ireland. It is very distinct from the Tramore form in being very small (valve lengths of 7.5–11.0 mm) and by having a ventral muscle scar 94% as long as wide compared to 160% in the Tramore form. They are similar in the subtriangular outline and the anterior extension of the ventral muscle scar being 44% as long as the valve length in *G. minor*.

Glyptambonites musculosus was described by Cooper (1956) from the Oranda Formation, lower Sandbian, Virginia, USA. It can be distinguished from the Tramore form in having a broadly rounded anterior margin and a subrectangular ventral muscle scar with almost parallel lateral margins and muscle bounding ridges that do not converge on the anterior end of the muscle scar. The anterior adductor scar is also distinct being oval and the long diductors are not depressed as in *Glyptambonites* sp.

Occurrence. – Tramore Limestone Formation, Quillia Section 2.

Family Grorudiidae Cocks & Rong, 1989

Genus *Grorudia* Spjeldnæs, 1957

Type species (by original designation). – *Grorudia grorudi* Spjeldnæs, 1957 from the Vollen Formation, lowermost Sandbian (*TS.*5a), Tåsen Station, Oslo-Asker District, Norway.

Grorudia grorudi Spjeldnæs, 1957

Plate 3, figures 15–20; Plate 4, figures 1–2, 4

1957 *Grorudia grorudi* Spjeldnæs, p. 62, pl. 1, figs 7, 10–11, text-figs 11D, 16.

1965 *Grorudia grorudi* Spjeldnæs; Williams *in* Muir-Wood & Williams, p. H373, figs 1a, b, p. H374.

1989 *Grorudia grorudi* (Spjeldnæs); Cocks & Rong, p. 112 *pars*, figs 64a, b, *non* figs 65a, b, 66, 67.

2000   *Grorudia grorudi* (Spjeldnæs); Cocks & Rong, p. 327 *pars*, fig. 213 (3b, c), *non* fig. 213 (3a).

2008   *Grorudia grorudi* (Spjeldnæs); Hints & Harper, pp. 275–276.

Comments on the synonymy list. – Cocks & Rong (1989, 2000) suppressed *Alwynella* as a junior synonym of *Grorudia* but Hints & Harper (2008), separated the two genera based on their distinct internal and external characters, which is followed here. The specimens illustrated by Cocks & Rong (1989, figs 65a, b, 66, 67) and Cocks & Rong (2000, fig. 213 (3a)) belong to the genus *Alwynella*.

Material. – 32 ventral valves and 13 dorsal valves.

Description. – Gently biconvex to planoconvex, transversely rhomboidal *Grorudia* species, 2–6 mm long ($n = 42$) and 5–20 mm wide ($n = 41$) with maximum width along straight hingeline and alate cardinal angles; ventral valve 37% as long as wide (24–55%, $n = 31$) and dorsal valve 31% as long as wide (20–35%, $n = 10$); ventral and dorsal interareas short, subequal, apsacline and catacline, respectively; ornament paucicostellate, differentiated with every third to fifth costa being stronger.

Ventral interior with no clear differentiation of muscle scars, muscle scar triangular and short, divided posteriorly in some specimens by a very short median septum; prominent, divergent mantle canals branch at 65% the length of the ventral valve anterior to the umbo (45–97%, $n = 11$); small, double teeth not readily seen; hingeline not denticulated.

Dorsal interior with cardinalia consisting of a stout, trifid cardinal process and short, widely divergent socket ridges; double sockets well-developed; subcircular adductor scars deeply impressed on a short, bilobed bema, adductor scar 70% as long as wide (50–86%, $n = 13$) and 55% as long as the valve length (41–73%, $n = 10$), bisected by equidistributate mantle canals and a median septum which extends for a short distance anterior to the muscle scar.

Discussion. – The specimens described here, were assigned to the genus *Grorudia* rather than *Alwynella* based on their lack of denticulation along the hingeline, their trifid cardinal process, which was not undercut, and their bilobed bema.

Spjeldnæs (1957) erected the genus *Grorudia* as well as the two species *G. grorudi* and *Grorudia glabrata* from his material collected in the Vollen Fm, Oslo District, Norway. The *Grorudia* specimens

from this study are similar to *G. grorudi* and are regarded as conspecific with this. Spjeldnæs (1957, p. 62) mentioned a species of *Grorudia* found in the Tramore Limestone which is 'extremely wide (hingeline/length ratio more than 4)' compared to *G. grorudi*. This study observed only three specimens with a width:length ratio this large (corresponding to a valve length:width ratio of 20–25%) and as they are similar to the other specimens in all other aspects they are here considered as a natural variation within *G. grorudi*.

Grorudia grorudi is distinguished from *G. glabrata* on the size, shape and ornamentation, *G. glabrata* being larger with a valve length:width ratio of 67%. They both have paucicostellate ornamentation but in *G. glabrata* the stronger costae are separated by six to seven fine costellae compared to two to four in *G. grorudi*. They can also be distinguished on the bema shape, which is bilobed in *G. grorudi* and circular in *G. glabrata*.

Since the erection of the genus *Grorudia* in 1957, only one new species, *Grorudia morrisoni* Hints & Harper, 2008, has been assigned to it. *Grorudia morrisoni* was described from the uppermost Darriwilian? (upper *TS.4c*) and middle Sandbian (*TS.5b*) in west Latvia. *Grorudia grorudi* can be distinguished from *G. morrisoni* by its ornament, which in *G. morrisoni* is paucicostellate to multicostellate with two orders of coarser costae separated by up to 12 delicate, weakly developed costellae and also on the presence of rugae in *G. morrisoni*. Their outlines are also different being rhomboidal in *G. grorudi* and subcircular in *G. morrisoni* and they differ in the course of the ventral mantle canals, which are divergent in the former and subparallel in the latter.

Occurrence. – Tramore Limestone Formation, Units 3–5, Barrel Strand area and Quillia Section 2. Dunabrattin Limestone Formation, Units III–IV, Dunabrattin–Bunmahon area.

Family Leptestiidae Öpik, 1933a; emend. Cocks & Rong, 1989

Genus *Leptestiina* Öpik, 1933a

Type species (by original designation). – *Benignites (Leptestiina) prantli* Havlíček, 1952, from the Králův Dvůr Formation, lower Ashgill (*TS.6b*), Bohemia, Czech Republic.

Remarks. – Cocks & Rong (2000) suppressed *Leptestiina* as a subgenus of *Leangella* Öpik, 1933a;

but we follow Candela & Harper (2010) in retaining *Leptestiina* as a separate genus based on the lack of an anterior platform in the early members of the *Leptestiina-Tufoleptina* (= *Leangella*) phylogenetic sequence of Mélou (1971) (including *L. derfelensis*) as described by Candela & Harper (2010).

Leptestiina derfelensis (Jones, 1928)

Plate 4, figures 3, 5–9

1928 *Leptelloidea derfelensis* Jones, p. 479, pl. 25, figs 3–7.

1955 *Leptellina derfelensis* (Jones); Williams *in* Whittington & Williams, p. 415, pl. 39, figs 71–73.

1968 *Leptestiina derfelensis* (Jones); Bates, p. 171, pl. 9, figs 7–9.

1971 *Leptestiina derfelensis* (Jones); Mélou, p. 99, pl. 2, figs 4–9.

1978 *Leptestiina derfelensis* (Jones); Cocks, p. 94.

1989 *Leangella (Leptestiina) derfelensis* (Jones); Cocks & Rong, p. 116.

2008 *Leangella (Leptestiina) derfelensis* (Jones); Cocks, p. 84.

2010 *Leangella (Leptestiina) derfelensis derfelensis* (Jones); Cocks, p. 1185, pl. 7, fig. 8.

Material. – Total of 28 ventral valves and eight dorsal valves.

Description. – Transverse to subcircular, concavo-convex *Leptestiina* species, 3–9 mm long ($n = 34$) and 6–13 mm wide ($n = 36$) with maximum width along straight hingeline; cardinal angles acute but commonly close to 90°; lateral and anterior margins smoothly curved; ventral valve moderately and evenly convex, 53% as long as wide (39–71%, $n = 28$) and 32% as deep as long (22–44%, $n = 25$); dorsal valve smoothly convex, 52% as long as wide (43–63%, $n = 6$) and 20% as deep as long (15–24%, $n = 2$); ventral interarea apsacline, delthyrium partly closed with apical pseudodeltidium; ornament finely unequiparvicostellate with prominent and fine costellae numbering two and ten per mm, respectively, at 5 mm anterior to the umbo.

Ventral interior with wide subpentagonal muscle scar, 58% as long as wide (39–88%, $n = 26$) and 21% as long as the maximum valve length (16–29%,

$n = 26$); short, narrow adductor scars lie between two larger, triangular diductors, the former extending further posteriorly and not as far anteriorly as the latter; teeth short and stout; the interiors of young valves usually bear marked radial papillae, which may be concentrated in an arc about the centre of the valve and around the margins; mantle canal system poorly impressed but two main submedian canals are observed extending from the anterior extremities of the diductor scars.

Dorsal interior with wide, trifid cardinal process ankylosed anteriorly to the valve floor; socket ridges widely divergent, flat lying and bladelike, almost parallel with hingeline; cardinalia 35% as long as wide (26–50%, $n = 6$) and 14% as long as the valve length (11–17%, $n = 5$); bema bilobed, striated and anteriorly undercut, with a moderate median indentation and splaying and bending sharply posterolaterally; 65% as long as wide (58–83%, $n = 7$) and 62% as long as the valve length (59–64%, $n = 5$); mantle canal systems not impressed.

Discussion. – The described specimens from the Tramore and Dunabrattin limestones are regarded as conspecific with *L. derfelensis* (Jones, 1928) from the Derfel Limestone Formation, upper Sandbian (lowermost *TS*.5b–lower *TS*.5c), Derfel Gorge, Gwynedd, North Wales. Jones (1928, p. 479) provided lengths and widths for nine mature specimens (lengths of 6.5–8.2 mm) and it is evident that the larger specimens from Tramore (lengths of 5.6–9.3 mm) have similar valve length:width ratios, Jones' specimens being 58% as long as wide respectively (53–66%) compared to 59% (49–71%) for seven Tramore specimens. They are also similar in outline, rib counts and dorsal interiors including the anterior extension of the bema being 65% as long as the valve length measured on two of Jones' illustrated specimens (1928, pl. 25, figs 3, 4) compared to 62% for the Tramore form.

Leptestiina oepiki oepiki (Whittington, 1938) from the upper Sandbian (lowermost *TS*.5b–lower *TS*.5c) of Powys, Wales, is morphologically closely related to *L. derfelensis*. Williams (1963) emended the diagnosis of *L. oepiki oepiki* and suggested this species and *L. derfelensis* may be synonymous. He noted that *L. oepiki oepiki* differs from *L. derfelensis* in convexity, ornamentation and in the shape of the dorsal bema but that better preserved samples were needed to assess whether these differences were significant. It would appear from his plates that the bema of *L. oepiki oepiki* is more deeply undercut than that of *L. derfelensis*. From Williams' measurements, it is evident that the dorsal bema is longer relative to width (less transverse) in *L. oepiki oepiki* than in *L. derfelensis* (76% and 65%, respectively), the ventral valve of *L. oepiki oepiki* is deeper than that of *L. derfelensis* (44% and 32%, respectively), and the ventral muscle scar is notably longer relative to its width (75% and 58%, respectively) and also longer relative to the ventral valve length (31% and 21%, respectively). *Leptestiina oepiki oepiki* also has a deeply impressed ventral vascular system not observed for *L. derfelensis*. Cocks (2010) revised *L. derfelensis* and *L. oepiki oepiki* and erected the subspecies *L. derfelensis derfelensis* and *L. derfelensis oepiki*. He argued that they are very similar but can be distinguished by the relatively wider bema in *L. derfelensis oepiki*. The above discussion indicates that they are probably significantly more different than described by Cocks (2010) and the separation of them in two species is retained here.

Leptestiina oepiki ampla Parkes, 1994; was described from the Annestown Formation, upper Sandbian (lowermost *TS*.5b–lower *TS*.5c), Carrigadaggan, Co. Wexford, southeast Ireland. It appears to be similar to *L. oepiki oepiki* in many aspects. It differs most significantly from *L. oepiki oepiki* and *L. derfelensis* by its notably more transverse bema, the length being 46% of the width for two specimens compared to 63–70% for four specimens of *L. oepiki oepiki* and 65% for seven specimens of *L. derfelensis* from this study. Cocks (2010, p. 1186) suggested that *L. oepiki ampla* may be a junior synonym of *derfelensis* because of its 'only slightly more transverse bema' but that more material of *ampla* was needed to verify this. We suggest retaining *derfelensis* and *oepiki ampla* as two separate species as this study indicate a significant difference between their bema length: width ratios. More material is needed to confirm this statistically, though.

Occurrence. – Tramore Limestone Formation, Unit 3 and 5, Barrel Strand area. Dunabrattin Limestone Formation, Unit II and IV, Dunabrattin–Bunmahon area.

Family Sowerbyellidae Öpik, 1930

Subfamily Sowerbyellinae Öpik, 1930

Genus *Sowerbyella* Jones, 1928

Subgenus *Sowerbyella* (*Sowerbyella*) Jones, 1928

Type species (by original designation). – *Leptaena sericea* J. de C. Sowerby, 1839, from the Alternata

Limestone Formation, lowermost Katian (middle *TS*.5c), Whittingslow, Shropshire, England.

Sowerbyella (*Sowerbyella*) *antiqua* Jones, 1928

Plate 4, figures 10–17

1852	*Leptaena tenuissimestriata* (M'Coy) M'Coy in Sedgwick & M'Coy, p. 239, pl. 1H, fig. 44.
1928	*Sowerbyella antiqua* Jones, p. 419, pl. 21, figs 7–11.
1949b	*Sowerbyella antiqua* var. *llandeiloensis* Williams, p. 234, pl. 11, figs 12–14.
1974	*Sowerbyella antiqua* Jones; Williams, p. 130, pl. 22, figs 4, 7–14, pl. 23, figs 1, 3, 4.
1974	*Sowerbyella* cf. *antiqua* Jones; Williams, p. 131, pl. 23, figs 2, 5–13.
1978	*Sowerbyella antiqua* Jones; Cocks, p. 97.
1981	*Sowerbyella antiqua* Jones; Lockley & Williams, p. 58, figs 196–212.
1989	*Sowerbyella* (*Sowerbyella*) *antiqua* Jones; Cocks & Rong, p. 139.
2008	*Sowerbyella* (*Sowerbyella*) *antiqua* Jones; Cocks, pp. 93–94.

Material. – 37 ventral valves plus eight small fragments and 35 dorsal valves plus 15 small fragments.

Description. – Concavoconvex *Sowerbyella* (*Sowerbyella*) species with outline shape changing through ontogeny from very transverse in young specimens to subcircular in mature individuals, 1–6 mm long ($n = 70$) and 2–11 mm wide ($n = 68$) with maximum width along straight hingeline and lateral and anterior margins gently rounded; cardinal angles commonly acute and sometimes slightly alate; ventral valve gently convex, 51% as long as wide (34–74%, $n = 36$) and 23% as deep as long (13–38%, $n = 20$); dorsal valve planar to gently concave, 52% as long as wide (34–75%, $n = 30$); ventral and dorsal interareas apsacline and anacline respectively with a small apical pseudodeltidium and complementary chilidium; ornament finely parvicostellate with counts of 11 ribs per mm at 2 mm anteromedially to the umbo (9–13, $n = 12$), two of seven valves show unequiparvicostellate ornament but none have rugae.

Ventral interior with small teeth and poorly developed dental plates; ventral muscle scar deeply impressed and strongly cordate, 69% as long as wide

(56–95%, *n* = 13) and 26% as long as the valve length (18–44%, *n* = 13); diductor scars diverge anteriorly and each may be divided by strong *vascular media*; adductor scars small, bilobed, impressed in the umbonal area and divided by a fine median ridge; mantle canal system lemniscate.

Dorsal interior with widely divergent, trifid cardinal process undercut anteriorly and fused with chilidial plates; socket plates straight, 29% as long as their lateral spread (21–57%, *n* = 30) and 10% as long as the valve length (8–18%, *n* = 30); a weak notothyrial platform may be developed; lophophore and muscle supports consisting of a median septum and three pairs of lateral septa radiating from the umbo and resting directly on the valve floor in young specimens but are ankylosed to form a raised bema in older specimens; median septum extending anteriorly for 52% (30–69%, *n* = 30) of the valve length measured from the umbo; lateral septa may extend a little longer anteriorly than the median septum; mantle canal system not impressed.

Remarks on the material. – Sixty three of the 72 specimens measured in this study for the taxonomical analyses were observed on a single 19 × 8 cm rock surface, likely collected in upper Unit 2, and comprising a layered coquina consisting of several shell beds with imprints of primarily *Sowerbyella* (*Sowerbyella*) valves (Fig. 18). This fossiliferous sample, as well as the remainder of the samples, consisted of calcareous siltstone indicating a water depth of deeper DR 3 between fair weather wave base and storm wave base (Fig. 5). A minor proportion of the valve imprints on the rock surface represents complete or almost complete valves of good preservational state and the remainder consists of smaller poorly preserved fragments. The orientations of the *Sowerbyella* (*Sowerbyella*) valves on the rock surface were analysed in PAST with chi-square directional statistics and the valve lengths were tested for normal distribution to assess if the assemblage represented an *in situ* population or if it was transported. The original orientation of the slab is unknown; hence, an artificial north (0°) was applied for the directional analysis.

The valve lengths measured on the slab were normally distributed (Shapiro–Wilk *P* = 0.96, significance level = 0.05) and the valve orientations on the rock surface show a preference for the 45–225° direction, that is northeast–southwest, for 60 specimens (Fig. 19). The preferred orientation of the valves is statistically significant with a *P*(same) value of 0.02 from the chi-square test at significance level 0.05. The normally distributed population with oriented valves indicate the analysed shell bed surface

represents a storm layer consisting of a winnowed lag formed during peak storm conditions (Kreisa & Bambach 1982).

The total assemblage of *Sowerbyella* (*Sowerbyella*) valves including both the valves from the rock surface and the other samples have normally distributed valve lengths when analysed together (Shapiro–Wilk *P* = 0.98). Hence, they were treated as belonging to the same population in the following statistical analyses.

Morphological variation. – The immature (small) and mature (large) specimens of *Sowerbyella* (*Sowerbyella*) *antiqua* from the Tramore area are notably different in the development of their dorsal muscle and lophophore supports and could easily be mistaken for two different species. Two distinct forms of dorsal valves were observed: one type with muscle and lophophore supports in the form of a median septum and three pairs of lateral septa resting directly on the valve floor (see Pl. 4, figs 10, 13), and one type with a distinct but variably developed, raised bema containing a median septum and three pairs of lateral septa (see Pl. 4, figs 11, 14 for moderately raised bema and Pl. 4, figs 12, 15 for highly raised bema). The development of a bema in later ontogenetic stages of *S.* (*S.*) *antiqua* was also observed by Williams (1974) and Lockley & Williams (1981) but is confirmed here by univariate and bivariate statistics and the multivariate principal components analysis (PCA) on the correlation matrix. The correlation matrix was applied as the analysed variables are in different units.

The univariate, bivariate and multivariate analyses of the dorsal valves included 25 specimens. The analysed variables include dorsal valve length (*L*_valve) and width (*W*_valve) and presence/absence of bema. The dorsal valves were assigned a number according to bema development: valves with septa only and no bema were assigned the value 1 in the PCA and illustrated as squares in Figure 20; and valves with a moderately to highly raised bema were assigned the value 2 and illustrated as circles in Figure 20. Describing the precise development of bema is very subjective because of its small size and the varying preservational stage of the dorsal valves ranging from very good to very eroded. Hence, it was not attempted to assign specimens with a moderately developed bema to a category of their own and these were included in the group with strongly developed bema.

The PCA clearly indicates a differentiation between the smaller and larger specimens (Fig. 20). Bivariate statistics confirm the significance of this difference showing that the mean valve length of the specimens without bema versus specimens with

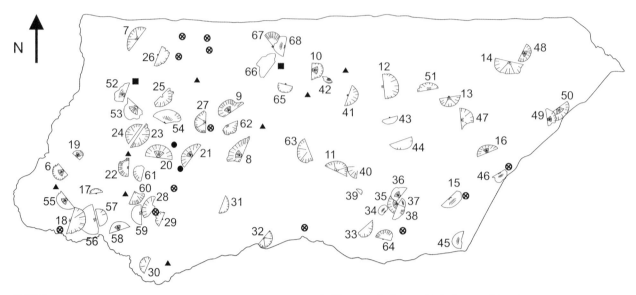

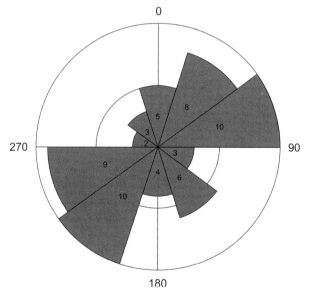

Fig. 18. Rock surface containing 63 measured valves of *Sowerbyella (Sowerbyella) antiqua*. Specimens are shown with standardized features such as dorsal bema and ventral muscle scar. The outline shapes are illustrated approximately as they were observed on the rock surface. Valve sizes and other measurements are provided in Appendix B. The maximum length and width of the rock is 19 × 8 cm.

bema is significantly different and that the bema is typically present in the large individuals (P(same) for $t < 0.05$; see Table 9). There is some overlap in the valve lengths of the two groups, the group without bema ranging in length from 2.1 to 3.0 mm and the group with bema from 2.2 to 4.8 mm, however; the latter only includes three valves, out of 18, of less than 3.1 mm.

The dorsal valves analysed here belong to different growth stages of *S. (S.) antiqua*. Specimens with septa only and no bema represent the younger immature forms of the assemblage and specimens with strongly developed bema represent older mature forms. The specimens with moderately developed bema probably represent growth stages in between the immature and mature forms but this was not investigated here. If the two different dorsal valve types belonged to two different species, they should each include a wider range of valve sizes representing the different ontogenetic stages of their individual species.

Discussion. – The genus *Sowerbyella* was originally described by Jones (1928) from the Llandeilo beds, uppermost Darriwilian (upper *TS*.4c), Dyfed, Wales, and later revised by Lockley & Williams (1981). The *Sowerbyella (Sowerbyella)* specimens from this study are similar to *S. (S.) antiqua* Jones, 1928; and are regarded conspecific. *Sowerbyella (Sowerbyella) antiqua* has previously only been described from Wales

Fig. 19. Rose diagram of the hingeline orientations of 60 *Sowerbyella (Sowerbyella) antiqua* valves measured on the 19 × 8 cm rock surface. 0° is the artificial north. There is a higher frequency of specimens oriented northeast–southwest (45–225°). The preferred orientation is statistically significant: chi-square = 9.73, P (same) = 0.02, significance level = 0.05.

and Shropshire, but its presence is now also confirmed from the Tramore area of southeast Ireland. *Sowerbyella (Sowerbyella)* cf. *S. (S.) antiqua* Jones was described by Williams (1974) from the Meadowtown Beds, uppermost Darriwilian (upper *TS*.4c), near Meadowtown, Shropshire, England. The large number of specimens of *S. (S.) antiqua* at hand, which includes almost the entire range of growth stages, demonstrates that *S. (S.)* cf. *antiqua sensu* Williams (1974) is conspecific with *S. (S.) antiqua*. The slight differences in ventral valve depth:length ratio and ventral muscle scar length:valve length ratio observed by Williams is not significant.

Sowerbyella (Sowerbyella) antiqua is very similar to *Sowerbyella (Sowerbyella) multipartita* Williams *in* Cocks, 1978. This species was described from the Spy Wood Grit Formation, lower Sandbian (*TS.*5a–lowermost *TS.*5b), near Rorrington, Shelve Inlier, Shropshire, England. *Sowerbyella (Sowerbyella) antiqua* and *S. (S.) multipartita* are similar in valve size and all valve ratios, except one as shown by Williams (1974, pp. 132, 133), and the development of a dorsal bema in larger specimens of *S. (S.) antiqua*. They only have slightly different rib counts with 10 ribs per mm at 2 mm anteromedially to the umbo in *S. (S.) multipartita* and 11 in *S. (S.) antiqua*. *Sowerbyella (Sowerbyella) multipartita* is, apart from *S. (S.) antiqua*, the only of the British species with three pairs of lateral septa. They can be distinguished primarily based on the arrangement of the dorsal valve septa; the septa of *S.*

(S.) multipartita extends anteriorly for 73% of the valve length for a population including most growth stages (Williams 1974) compared to an average of 52% for the total Tramore population. The median septum of *S. (S.) multipartita* furthermore extends beyond the lateral septa, the median septum of *S. (S.) antiqua* terminating before these. The occasionally developed dorsal median fold in *S. (S.) multipartita* extending from the protegulum and fading out anteriorly (Williams 1974) was not observed in *S. (S.) antiqua*.

Sowerbyella (Sowerbyella) antiqua is most closely related to the other Ordovician British species of *Sowerbyella (Sowerbyella)* and a brief comparison focusing on the most distinct differences is provided here: Of the fourteen formally and one informally described species mentioned by Cocks (2008) (excl. *S. (S.) multipartita*), *S. (S.) antiqua* is most readily distinguished from all of them by the number of lateral septa developed in the dorsal valve, *S. (S.) antiqua* having three pairs and all the other species having fewer pairs. *Sowerbyella (Sowerbyella) elusa* Williams, 1962; *S. (S.) fallax* Jones, 1928; *S. (S.)* cf. *S. (S.) monilifera* Cooper, 1956; *S. (S.) musculosa* Williams, 1963; *S. (S.) sericea* J. de C. Sowerby, 1839; *S. (S.) sericea permixta*

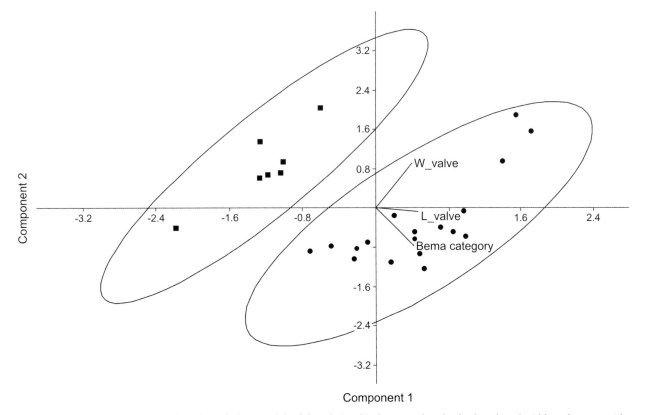

Fig. 20. Principal components analysis (correlation matrix) of the relationship between dorsal valve length and width and presence/absence of bema in *Sowerbyella (Sowerbyella) antiqua*. L_valve, valve length; W_valve, valve width; squares, valves with septa and no bema; circles, valves with moderately to highly raised bema. Large circles are 95% ellipses. Variances: principal component 1 = 63.9%; principal component 2 = 23.0%; principal component 3 = 13.8%.

Table 9. *Sowerbyella* (*Sowerbyella*) *antiqua*; univariate and bivariate statistics for dorsal valves with and without bema.

	N	Mean (mm)	Shapiro–Wilk (normality)	F	P(same) F	Student's t	P(same) t
Bema absent	7	2.8	0.84	3.49	0.13	−3.18	4×10^{-3}
Bema present	18	3.8	0.92				

Significance level: 0.05.

Williams, 1963; *S.* (*S.*) *soudleyensis* Jones, 1928; *S.* (*S.*) *thraivensis* (Reed, 1917) and *S.* (*S.*) sp. (Mitchell 1977) only have one submedian pair of septa. *Sowerbyella* (*Sowerbyella*) *sladensis* Jones, 1928; and *S.* (*S.*) cf. *varicostellata* Cooper, 1956 (also described by Williams 1962, p. 180) have two pairs of lateral septa one of them submedian. All the above mentioned species can furthermore be distinguished from *S.* (*S.*) *antiqua* based on individual morphological characters but they also display some similarities with this species, the most common being the subcircular outline, the unequally parvicostellate nature of the external ornamentation and the concavoconvex profile.

Occurrence. – Tramore Limestone Formation: Units 2–5, Barrel Strand area, Pickardstown section and Quillia sections 1 and 2. Dunabrattin Limestone Formation, Unit II and IV, Dunabrattin–Bunmahon area.

Order Billingsellida Schuchert, 1893

Suborder Clitambonitidina Öpik, 1934

Superfamily Clitambonitoidea Winchell & Schuchert, 1893

Family Clitambonitidae Winchell & Schuchert, 1893

Genus *Atelelasma* Cooper, 1956

Type species (by original designation). – *Atelelasma perfectum* Cooper, 1956, from the Arline Formation, upper Darriwilian (*TS.*4c), Tennessee, USA.

Atelelasma longisulcum n. sp.

Plate 4, figures 18–21; Plate 5, figures 1–2, 5–6;
Plate 10, figure 3

Derivation of name. – Alluding to the long dorsal sulcus of the species.

Holotype. – NHMUK PI BB35204, interior mould of dorsal valve, Tramore or Dunabrattin Limestone Formation (Pl. 4, figs 19, 20).

Material. – Total of 10 ventral valves and four dorsal valves.

Diagnosis. – Subcircular to subquadrate *Atelelasma* species with maximum width anterior of hingeline; ventribiconvex with evenly convex, sulcate dorsal valve with sulcus extending from the umbo to the anterior margin; ventral interarea catacline to strongly apsacline; dorsal interarea anacline; ornament variably lamellose and moderate finely costellate with counts of three and four costellae per mm at 5 mm anterior to the umbo; ventral interior with shallow spondylium simplex and weak or absent median septum; dorsal interior with simple, fine cardinal process and a well-developed notothyrial platform.

Description. – Ventribiconvex, subcircular to subquadrate, *Atelelasma* species with outline changing from transversely wider than long to equidimensional through ontogeny; lateral and anterior margins gently rounded; cardinal angles usually somewhat wider than 90° but slightly acute in one specimen; 4–22 mm long ($n = 13$) and 5–21 mm wide ($n = 13$) with maximum width anterior of hingeline; ventral hinge width 93% of maximum valve width (86–98%, $n = 8$), ventral valve 86% as long as maximum width (66–102%, $n = 9$) and 29% as deep as long (23–32%, $n = 5$); dorsal valve evenly convex; well-developed dorsal sulcus originates at the umbo, becomes wider and shallower anteriorly and extends to the anterior margin; dorsal valve 76% as long as wide (66–92%, $n = 4$) and 13% as deep as long (11–14%, $n = 2$); ventral interarea long, catacline in small specimens and strongly apsacline in larger specimens (Pl. 10, fig. 6), 29% as long as the valve length (21–33%, $n = 7$); delthyrium open but with a slight development of deltidial plates laterally; dorsal interarea shorter, anacline, chilidium not clearly seen; ornament moderate finely costellate with aditicules and occasionally swollen costellae, ribs arising by intercalation and counts of three and four costellae per mm at 5 mm anterior to the umbo for one and two dorsal valves, respectively; concentric ornament of lamellae may be strongly developed, especially peripherally.
Ventral interior (Pl. 10, fig. 3) with shallow spondylium simplex, 90% as long as wide (75–125%,

$n = 9$) and 33% as long as the valve length (26–47%, $n = 8$); muscle scars rarely well-developed, spondylium simplex supported by a median septum in four of nine valves; teeth simple, not strongly developed; mantle canal system pinnate but weakly impressed, vascula media extend anteriorly on either side of the median septum with weak impressions of vascula genitalia posterolaterally.

Dorsal interior with short, fine, ridgelike cardinal process; notothyrial platform well-developed, passing anteriorly into a broad median ridge which extends to the base of the dorsal adductor scars; brachiophores widely divergent, massive, fused with the notothyrial platform and may be slightly undercut posteriorly by the sockets; dorsal adductor scars quadripartite and deeply impressed, 68% as long as wide (58–78%, $n = 2$) and 47% as long as the valve length (46–47%, $n = 2$), divided medially by the broad median ridge and laterally by vascula myaria; median ridge 47% as long as the valve length ($n = 1$); mantle canal system weakly impressed.

Morphological variation. – Atelelasma longisulcum n. sp. was investigated for allometric growth. The valve length and width (L_valve and W_valve, respectively) included a sufficient number of measurements for statistical analysis.

The log-transformed RMA plot (Fig. 21) of the valve widths plotted against valve length indicates a possible breakpoint between valve lengths of 16.0–16.5 mm (1.26–1.31 on the log-transformed plot): valve lengths of 3.5–16.0 mm mainly plot above the regression line, representing relatively wider valves compared to valve length (squares in Fig. 21) and valve lengths of 16.5–21.5 mm plot beneath this line, representing relatively less wide valves (circles in Fig. 21). The breakpoint between valve lengths of 16.0 and 16.5 mm indicates that a change in the valve length–valve width growth rate possibly took place at this growth stage. Unfortunately the amount of material was not sufficient to divide the data into two groups of smaller and larger valves, respectively, and compare the slopes of their log-transformed regression lines bivariately to assess the growth rate changes. This type of analysis was attempted but the bootstrapping in PAST failed for the group of larger valves.

The log-transformed RMA of the total assemblage indicates allometric growth as $P(a = 1) < 0.05$; that is, the relative growth rate between the valve length and width changed through ontogeny (Fig. 21). The growth patterns are most easily illustrated in a non-transformed RMA of the valve lengths plotted against the valve length:width ratios

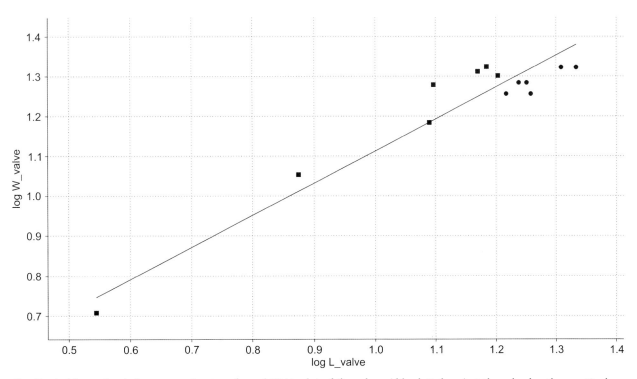

Fig. 21. Atelelasma longisulcum n. sp. Log-transformed RMA plot of the valve width plotted against the valve length. $n = 13$, slope $a = 0.81$, error $a = 0.07$, $r = 0.96$, P(uncorr) $= 3 \times 10^{-7}$, $P(a = 1) = 0.02$, significance level $= 0.05$. The RMA indicates statistically significant allometry as $P(a = 1) < 0.05$.

(L_valve × 100/W_valve) (Fig. 22). The length:
width ratio increases considerably with increased
valve length showing that the valve outline evolved
from transverse in the smaller, immature specimens,
to equidimensional in the large, mature specimens.
It is also evident from Figure 22 that the valve width
growth rate experienced a sudden increase relative to
valve length at the growth stage represented by the
breakpoint (valve lengths of 16.0–16.5 mm).

Discussion. – This is the first record of *Atelelasma* in
southeast Ireland. Only two confirmed occurrences
of *Atelelasma* have previously been recorded from
the British Isles: *Atelelasma anatolicum* MacGregor,
1961; from Wales and *Atelelasma* sp. described by
Mitchell (1977) from North Ireland.

Atelelasma longisulcum n. sp. is very similar to
Atelelasma sp. described by Mitchell (1977) from the
Bardahessiagh Formation, upper Sandbian (lowermost *TS*.5b–middle *TS*.5c), Pomeroy, Co. Tyrone,
North Ireland. They have similar dorsal rib counts
A. longisulcum n. sp. having 3–4 ribs per mm at
5 mm anterior to the umbo compared to three in
Atelelasma sp. They both have a sulcus extending
from the umbo to the anterior margin and their
quadripartite dorsal adductor muscle scar is also
comparable being 68% as long as wide in

A. longisulcum n. sp. and 63% in *Atelelasma* sp. They
also have equally long dorsal median septa extending
from the notothyrial platform for 47% and 50% the
valve length, respectively. *Atelelasma longisulcum* n.
sp. and *Atelelasma* sp. differ in the length of the dorsal adductor scar relative to valve length being 47%
and 60%, respectively. The dorsal valve of *Atelelasma*
sp. furthermore includes a large, convex chilidium,
which is not observed in *A. longisulcum* n. sp. The
two Irish forms appear to be closely related but more
material of both species is needed for further comparison.

Atelelasma anatolicum was described by MacGregor (1961) from the Calcareous Ash Formation,
uppermost Darriwilian (upper *TS*.4c), Powys, Wales.
It is only slightly older than the Tramore species,
which probably appeared in the earliest Sandbian.
They are morphologically very different, one of the
most obvious differences being the folded dorsal
valve and sulcate ventral valve of *A. anatolicum*.

Of the twelve species of *Atelelasma* described and
figured by Cooper (1956), *A. longisulcum* n. sp. most
closely resembles *Atelelasma dorsoconvexum* Cooper,
1956; from the basal Athens Formation, upper Sandbian (upper *TS*.5b–lower *TS*.5c), Riceville, Tennessee, USA. Both forms are subcircular to
subquadrate with cardinal angles about 90°, they are

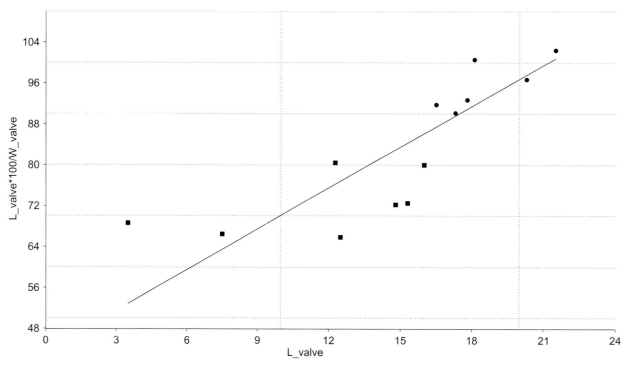

Fig. 22. Atelelasma longisulcum n. sp. RMA plot of the valve length:width ratio (L_valve × 100/W_valve) plotted against the valve
length. $n = 13$, slope $a = 2.66$, error $a = 0.47$, $r = 0.81$, P(uncorr) $= 7 × 10^{-4}$, significance level $= 0.05$. The length:width ratio increases
with increased valve length showing that the valve outline developed from transverse in the smaller specimens to equidimensional in the
large specimens.

ventribiconvex, have catacline to strongly apsacline ventral interareas and have equal mean rib counts of seven ribs per 2 mm at 5 mm anterior to the umbo with a range of 3–4 ribs. The dorsal sulcus of *A. dorsoconvexum* is shorter, however, and only extends from the umbo to the middle of the valve. They also differ in several size ratios with *A. dorsoconvexum* having a ventral hingeline width:valve width ratio of 83% (70–89%, *n* = 3), a spondylium length:width ratio of 120% (100–133, *n* = 3), a dorsal adductor scar length:width ratio of 78% (69–90%, *n* = 3) and a dorsal median septum length:valve length ratio of 31% (23–38%, *n* = 5). The compared values for Cooper's (1956) specimens were obtained by measuring his figured specimens (pl. 80, G, figs 35–39, pl. 82, M, figs 53–57).

Occurrence. – Tramore Limestone Formation, Unit 3, Barrel Strand area. Dunabrattin Limestone Formation, Units I–III, Dunabrattin–Bunmahon area.

Class Rhynchonellata Williams, Carlson & Brunton, 1996

Order Orthida Schuchert & Cooper, 1932

Suborder Orthidina Schuchert & Cooper, 1932

Superfamily Orthoidea Woodward, 1852

Family Orthidae Woodward, 1852

Genus *Sulevorthis* Jaanusson & Bassettt, 1993

Type species (by original designation). – *Orthis lyckholmiensis* Wysogórski, 1900; from the Upper Ordovician of Estonia. Neotype selected by Jaanusson & Bassett (1993) from the upper Katian (lower upper TS.5d), Kõrgessaare, Island of Hiiumaa, Estonia.

Discussion. – *Sulevorthis* aff. *S. blountensis* was assigned to the genus *Sulevorthis* following the original definition given by Jaanusson & Bassett (1993, pp. 37–38). They included forms with both bladelike and rodlike brachiophores as well as forms with or without a ventral median septum. Williams & Harper (2000, p. 724) revised *Sulevorthis* and *Orthambonites* and distinguished them, among other features, by the presence of bladelike brachiophores and a ventral median septum in *Sulevorthis*. They did not define the shape of the brachiophores in *Orthambonites* but commented that the genus lacks a ventral median septum. *Sulevorthis exopunctatus*

(Williams, 1974) possesses both rodlike brachiophores and a ventral median septum indicating such a distinction cannot be made and the current definition of the genera is in need of revision. This species also possesses, unusually, exopunctae.

Sulevorthis aff. *S. blountensis* (Cooper, 1956)

Plate 5, figures 3–4, 7–8, 11–12

aff. 1956 *Orthambonites blountensis* Cooper, p. 298, pl. 35, C, figs 26–34.

aff. 1993 *Sulevorthis blountensis* (Cooper); Jaanusson & Bassett, p. 38.

1994 'Orthambonites spp.' Parkes, p. 139 *pars*, pl. 3, figs 3–5, *non* figs 1, 2.

Comments on the synonymy list. – The specimens figured by Parkes (1994, pl. 3, figs 1, 2) appear to be more finely costellate than those of *Sulevorthis* aff. *S. blountensis*. They are poorly preserved though, and the costae are difficult to distinguish.

Material. – Total of 12 ventral valves and 10 dorsal valves.

Description. – Moderately transverse to subcircular, ventribiconvex *Sulevorthis* species with maximum width generally along the hingeline in immature specimens and posteromedially in adults; rounded cardinal angles vary from just acute to obtuse; margins gently rounded laterally becoming more strongly rounded anteriorly; 2–8 mm long (*N* = 22) and 3–10 mm wide (*n* = 21), ventral valve 76% as long as wide (66–86%, *n* = 11); moderately and evenly convex with a slightly incurved umbo, maximum convexity about posteromedially with flanks descending to flattened posterolateral margins; dorsal valve 73% as long as wide (62–83%, *n* = 10); gently convex and shallowly sulcate, the sulcus originating at the umbo and containing two costae with three or four costae on each slightly elevated flank; sulcus becomes weaker anteriorly and is indistinguishable at the anterior margin on mature specimens; ventral and dorsal interareas straight, apsacline and anacline respectively with open delthyrium and notothyrium; radial ornament costate with strong filae; ornamentation of 18–20 sharply rounded costae continuing distinctly and sharply to the beak; counts of 18, 19 and 20 ribs on 4, 1 and 1 valves respectively with a mode of 18; interspaces about equal in width to costae; wavelength of costae is 0.85 mm at 5 mm anterior to the umbo (*n* = 1).

Ventral interior with short, stout teeth which have well-developed elongate crural fossettes on their inner surfaces; dental plates extend for a short distance anteriorly and tend to converge; ventral muscle scar subtriangular, 74% as wide as long (62–88%, $n = 3$) and 32% as long as the valve length (27–36%, $n = 4$); details not seen due to poor preservation but the muscle scar does not seem to extend beyond the dental plates; *vascula myaria* obscure.

Dorsal interior with a simple ridgelike cardinal process in young specimens becoming greatly thickened in adults, especially posteriorly; rodlike brachiophores diverge from the well-developed notothyrial platform at about 90° and these too become massively thickened in adults; well-developed raised, cuplike sockets occur on the outer edges of the brachiophores; brachiophores 43% as long as their lateral spread (37–50%, $n = 6$) and 19% as long as the valve length (17–22%, $n = 6$); low, wide and rounded median ridge extends anteriorly from the notothyrial platform dying out medially in adult specimens; dorsal adductor scar quadripartite, deeply impressed posteriorly and more weakly impressed anteriorly with indistinct anterior margins; interior of valve margin deeply crenulated by impressions of costae, with shallower impressions of mantle canals between these; vascula myaria not impressed.

Discussion. – *Sulevorthis* aff. *S. blountensis* differs from most of the British *Sulevorthis* species (see Cocks 2008) by its rodlike brachiophores and from the American species (see Cooper 1956) as well at the British species by the very small size, the mode of 18 costae as well as various other morphological features.

The Tramore form is very similar to *S. blountensis* (Cooper, 1956) from the Arline Formation, upper Darriwilian (*TS*.4c), Virginia, USA. The specimens figured by Cooper (1956, pl. 35, C, figs 26–34) were measured for comparison. The Tramore and American forms are of the same small size, Cooper's specimens being 7–8 mm long ($n = 7$) and 9–10 mm wide ($n = 7$), with similar subcircular outline and acute to obtuse cardinal angles. They have comparable coarsely costate ornament with a mode of 18 ribs for both (*S. blountensis*: counts of 17 and 18 ribs on two and five valves, respectively). The dorsal sulcus of *S.* aff. *S. blountensis* is developed like that in *S. blountensis* with two costae in the shallow sulcus. The ventral interior of the two species are largely similar, both having stout teeth, well-developed crural fossettes, and a short ventral muscle scar 75% as wide as long ($n = 1$) and 39% as long as the valve length ($n = 1$) in *S. blountensis* compared to 74% (62–88%, $n = 3$) and 32% (27–36%, $n = 4$),

respectively, in *S.* aff. *S. blountensis*. They also both lack a ventral median septum. Their dorsal interiors differ somewhat: they both possess a ridgelike cardinal process, but this is thickened in adult specimens of *S.* aff. *S. blountensis*, which is not observed in *S. blountensis*. The dorsal median ridge is indistinct in *S. blountensis* and seems to be somewhat shorter compared to the valve length than that of *S.* aff. *S. blountensis* and the rodlike brachiophores of the former diverge at an angle of about 115° ($n = 1$) compared to about 90° in the latter. The collection of *S. blountensis* specimens described by Cooper (1956) needs to be reinvestigated and more and better preserved material of the Tramore species should be collected for a more precise assessment of the relation between these two forms.

Parkes (1994, pl. 3, figs 1–5) figured four interior moulds and one latex replica of '*Orthambonites*' spp. from the uppermost Sandbian (lower *TS*.5c) of Kildare and Kilbride, southeast Ireland. His specimens are not well preserved, but those figured on his plate 3, figures 3–5 are likely conspecific with *Sulevorthis* aff. *S. blountensis* as they are similar in size and seem to have completely similar valve shape, cardinal angles and ventral and dorsal interiors. It was not possible to count the ribs on Parkes' figured specimens but they seem about as coarsely costellate as the Tramore form.

Occurrence. – Tramore Limestone Formation, Units 1–4, Barrel Strand area.

Family Glyptorthidae Schuchert and Cooper, 1931

Genus *Glyptorthis* Foerste, 1914

Type species (by original designation). – *Orthis insculpta* Hall, 1847, from the Waynesville and Liberty formations, upper Katian (upper *TS*.5d), Ohio and Indiana, USA.

Glyptorthis crispa (M'Coy, 1846)

Plate 5, figures 9–10, 13, 15–20

1846 *Orthis crispa* M'Coy, p. 29, pl. 3, fig. 10.

1871 *Orthis crispa* M'Coy; Davidson, p. 256.

1871 *Orthis crispata* [*sic*] M'Coy; Davidson, pl. 38 *pars*, figs 5–8, ?fig. 10, *non* fig. 9.

1883   *Orthis crispa* M'Coy; Davidson, p. 176, p1. 13, fig. 7, 8.

non 1961   *Glyptorthis crispa* (M'Coy); MacGregor, p. 188.

1962   *Glyptorthis balclatchiensis* [*sic*] (Davidson); Williams, p. 109 *pars*, pl. 9, figs 26, 31, 32, *non* figs 21–25.

1964   *Glyptorthis crispa* (M'Coy); Wright, p. 173.

1974   *Glyptorthis crispa* (M'Coy); Williams, p. 65.

1978   *Glyptorthis crispa* (M'Coy); Cocks, p. 47.

2008   *Glyptorthis crispa* (M'Coy); Cocks, p. 118, pl. 5, fig. 21.

2014   *Orthis crispa* (M'Coy); Wright & Stigall, p. 894.

Comments on the synonymy list. – The dorsal interior of *Orthis crispata* figured by Davidson (1871, pl. 38, fig. 10) appears to have a slightly thicker cardinal process and better defined median septum than our specimens but this may not be significant. The specimen figured by Davidson (1871, pl. 38, fig. 9) does not belong to *G. crispa* as its ventral interiors are very different, for example by having adductor scars not prominently elevated from the valve floor and not extending forwards of the diductors. *Glyptorthis crispa* mentioned by MacGregor (1961, p. 188) in his comparisons of *Glyptorthis* species with his new species *G. minor* is not *G. crispa* as it has a strongly convex dorsal valve and lacks a dorsal sulcus. The specimens illustrated by Williams (1962, figs 26, 31, 32) probably belong to *G. crispa* as they have similar rib counts and dorsal interiors.

Material. – Total of 45 ventral and 43 dorsal valves together with one ventral and one dorsal valve moderately deformed by shearing but with well-defined interiors. The deformed valves were not included in the measurements for the taxonomical descriptions below but only used for studying the nature of the valve interiors.

Description. – Subrectangular to subquadrate, equibiconvex to dorsibiconvex *Glyptorthis* species with cardinal angles generally greater than 90°; 2–23 mm long ($n = 80$) and 3–26 mm wide ($n = 85$) with maximum width generally anterior to hingeline; hingeline 91% as wide as the valve width (76–100%, $n = 16$); ventral valve 82% as long as wide (56–98%,

$n = 40$) and 15% as deep as long (13–18%, $n = 2$) with gentle sulcus developed to varying degrees; dorsal valve 73% as long as wide (46–93%, $n = 40$) and 19% as deep as long (18–20%, $n = 2$) with a well-developed sulcus in young specimens flattening out between 6 and 9 mm and commonly replaced by a low fold producing a gently plicate anterior commissure; ventral interarea apsacline and gently curved, dorsal interarea short and anacline; ornament ramicostellate with a total of 14–22 rounded costae; secondary costellae commonly developed and tertiaries rarely developed; internal costellae are more commonly associated with costae 1–3 but internal and external costellae are about equally developed from costae 4–8; total number of ribs at 5 mm anterior to the umbo is 33–46 ($n = 25$) with a mode of 38; concentric lamellae strongly developed.

Ventral interior with suboval to subtriangular muscle scar, 88% as wide as long (63–142%, $n = 26$) and 36% as long as the valve length (23–47%, $n = 25$); anterior margins of the adductor scars convex to truncated at the front, indented at the junctions of the adductor and diductor scars; adductor scars extend slightly forwards of diductors and are prominently elevated above the floor of the valve; dental plates divergent, short, 21% as long as the valve length (16–29%, $n = 11$); mantle canal system saccate with strongly impressed *vascula media*; lateral branching from the median trunks of the *vascula media* occurs first at 63% the valve length anterior to the umbo ($n = 1$).

Dorsal interior with thin bladelike brachiophores divergent at about 90° to bound the notothyrium; cardinal process a simple, fine ridge extending to the anterior margin of the notothyrial platform; brachiophores 42% as long as their lateral spread (32–50%, $n = 14$) and 14% as long as the valve length (11–17%, $n = 14$); adductor scar quadripartite, poorly defined on smaller specimens and well-defined on large specimens, 74% as long as wide (59–98%, $n = 14$), and 36% as long as the valve length (25–51%, $n = 13$), bisected by a low, wide median ridge and a pair of thin lateral ridges, the former extending anteriorly from the notothyrial platform beyond the anterior margin of the adductor scar where it dies out; posterior adductor scars smaller than the anterior scars, subtriangular to subrectangular with their long axis oriented laterally; anterior scars subtriangular to subrectangular with their long axis oriented subparallel to the median septum; mantle canal system saccate with strongly impressed vascula media.

Discussion. – The *Glyptorthis* specimens described above are regarded as conspecific with *G. crispa*

(M'Coy, 1846). The original *G. crispa* material of M'Coy (1846) was primarily collected at Tramore, Co. Waterford, southeast Ireland, but a few specimens came from Pomeroy, North Ireland. Cocks (1978) selected a lectotype, which was probably collected in the Tramore area. The lectotype was figured by Cocks (2008, pl. 5, fig. 17). The specimens from Tramore illustrated by M'Coy (1846) and Cocks (2008) were measured for comparison with the specimens from this study. They are very similar with respect to hinge width:valve width ratio, valve length:valve width ratio and number of costae at 5 mm anterior of the umbo (Tables 10 and 11). From M'Coy's (1846) description of *G. crispa*, it is evident that it is similar to our specimens concerning its dorsibiconvex nature and its shallow dorsal sulcus. The specimen illustrated by M'Coy differs a bit from the average of our specimens and from the one figured by Cocks (2008) in being widest at the hingeline and by having cardinal angles slightly less than 90°, but these deviations fall within the observed variations of the species.

Wright (1964) discussed the history and development of *G. crispa* in connection with description of *Glyptorthis maritima maritima* Wright, 1964; from the Portrane Limestone Formation, middle upper Katian (lower middle *TS*.6a–lowermost *TS*.6b), east Ireland. Wright (1964, p. 173) proposed restricting *G. crispa* 'to the form occurring in the Tramore Limestone and the forms conspecific with it'. The Pomeroy specimens originally assigned to *G. crispa* by M'Coy (1846, p. 29) was

later reassigned to *Glyptorthis* cf. *G. concinnula* Cooper, 1956 by Mitchell (1977). That form was described from the Bardahessiagh Formation, upper Sandbian – lowermost Katian (middle *TS*.5b–middle *TS*.5c), Pomeroy, Co. Tyrone, Northern Ireland and by Candela & Harper (2010) from Wallace's Cast, Southern Uplands, Scotland. *Glyptorthis* cf. *G. concinnula* is clearly distinguished from *G. crispa* by its coarser ornamentation with rib counts of 19–22 at the 5 mm growth stage versus 33–46 in *G. crispa*. They can furthermore be distinguished based on the dorsal and ventral interiors, the dorsal cardinal process being a strong ridge in *G.* cf. *G. concinnula* compared to a simple fine ridge in *G. crispa*, and the ventral adductor scar being much wider compared to the total scar width in *G. crispa* than *G.* cf. *G. concinnula*.

Mitchell *et al.* (1972) recorded *Glyptorthis* sp. from the Courtown Limestone, Courtown, southeast Ireland, which is stratigraphically equivalent to the upper part of the Tramore Limestone Formation. This species was not described, however, and further details of ribbing patterns and statistical analyses on the morphology are needed for detailed comparison.

Elsewhere, Wright & Stigall (2014) have provided a careful and detailed, cladistic-based analysis of this disparate and species-rich genus across Laurentia. Unfortunately a similar analysis of the genus from the Avalonian and Ganderian terranes is premature and outside the scope of the current monograph. Nevertheless, three taxa (*G. glypta* Cooper, 1956;

Table 10. Comparison of *Glyptorthis crispa* from this study with the specimens of this species figured by M'Coy (1846, pl. 3, fig. 10) and Cocks (2008, pl. 5, fig. 21).

Character	Valve	Data source	*n*	Mean %	Range %
*W*_hinge × 100/*W*_valve	Ventral	This study	16	91	76–100
	Vent + dors	M'Coy (1846)	1	100	–
	Dorsal	Cocks (2008)	1	93	–
*L*_valve × 100/*W*_valve	Dorsal	This study	40	73	46–93
	Dorsal	M'Coy (1846)	1	68	–
	Dorsal	Cocks (2008)	1	71	–

L, length; *W*, width.

Table 11. Comparison of *Glyptorthis crispa* from this study with the specimens of this species figured by M'Coy (1846, pl. 3, fig. 10) and Cocks (2008, pl. 5, fig. 21).

Character	Valve	Data source	*n*	Mean	Range
Costae per mm at 5 mm anterior to the umbo	Dorsal	This study	25	38*	33–46
	Dorsal	M'Coy (1846)	1	~40	–
	Dorsal	Cocks (2008)	1	~40	–
Cardinal angles	Vent+dors	This study	80	>90	–
	Dorsal	M'Coy (1846)	1	<90	–
	Dorsal	Cocks (2008)	1	>90	–

*Mode.

G. senecta Cooper, 1956 and *G. sulcata* Cooper, 1956) are broadly coeval with *G. crispa*, and of these, *G. sulcata* appears most similar.

Occurrence. – Tramore Limestone Formation, Units 1–3, Barrel Strand area and Quillia Section 2. Dunabrattin Limestone Formation, Units I–III, Dunabrattin–Bunmahon area.

Family Hesperorthidae Schuchert & Cooper, 1931

Genus *Hesperorthis* Schuchert & Cooper, 1931

Type species (by original designation). – *Orthis tricenaria* Conrad, 1843, from the Guttenberg Formation, middle–upper Darriwilian (*TS.4b–TS.4c*), Wisconsin, USA.

Hesperorthis leinsterensis n. sp.

Plate 5, figure 14; Plate 6, figures 1–13; Plate 10, figures 7, 9

Derivation of name. – Alluding to the species' occurrence in the Leinster Terrane.

Holotype. – NHMUK PI BB35232, interior mould of dorsal valve, Tramore Limestone Formation (Pl. 5, fig. 14).

Material. – Total of 33 ventral valves and 20 dorsal valves.

Diagnosis. – Transversely suboval to subcircular, ventribiconvex *Hesperorthis* species; ventral valve strongly and evenly convex, 37% as deep as long; dorsal valve gently and evenly convex with a shallow sulcus becoming indistinguishable anteriorly; ventral interarea 16% as long as the valve length, strongly apsacline in adult specimens; dorsal interarea shorter, anacline; ornamentation of 24–25 rounded costae with three thin ridges in each interspace; dorsal valve with short brachiophores, 22% as long as the valve length; cardinal process a simple ridge.

Description. – Transversely suboval to subcircular, ventribiconvex *Hesperorthis* species with outline changing from wider than long to longer than wide through ontogeny; 2–15 mm long ($n = 53$) and 3–17 mm wide ($n = 53$); maximum width at or just anterior to the hingeline; cardinal angles orthogonal to obtuse; lateral and anterior margins

evenly rounded; ventral valve strongly and evenly convex, 81% as long as wide (56–115%, $n = 33$) and 37% as deep as long (29–46%, $n = 7$); dorsal valve gently and evenly convex 76% as long as wide (61–104%, $n = 20$) and 20% as deep as long ($n = 1$) with a faint shallow sulcus becoming indistinguishable anteriorly; ventral interarea on average 16% as long as the valve length (8–30%, $n = 17$) but changing through ontogeny with increasing interarea length relative to valve length and evolving from catacline to strongly apsacline and slightly curved (Pl. 10, fig. 9); dorsal interarea shorter, anacline; small apical pedicle collar may be developed; ornament costate with commonly 24 or 25 costae in total counted on four and nine valves, respectively; wavelength of costae at 5 mm anterior to the umbo varies from 0.7 to 0.8 mm with interspaces about equal in width to the costae and containing three fine ridges, the median one being the strongest; fine, concentric lamellae are observed cutting these.

Ventral interior with short, stout teeth (Pl. 10, fig. 9) and receding dental plates; ventral muscle scar elongately subtriangular, 87% as wide as long (57–138%, $n = 19$) and 32% as long as the valve length (25–41%, $n = 19$), muscle scar length increases relative to both muscle scar width and valve length through ontogeny with muscle scar shape evolving from wider than long to longer than wide; adductor scar weakly impressed, about one-third of the total scar width and not extending as far anteriorly as the diductor scars.

Dorsal interior (Pl. 10, fig. 7) with cardinalia composed of a fine, simple ridgelike cardinal process on a well-developed notothyrial platform; brachiophores short, divergent and wedge-shaped in cross-section, forming inner boundaries to well-developed, raised, cuplike sockets; brachiophores 76% as long as their lateral spread (65–86%, $n = 10$) and 22% as long as the valve length (19–25%, $n = 10$), the lateral spread of the brachiophores increasing relative to their length trough ontogeny; well-developed rounded median ridge extends anteriorly for 75% the length of the valve (71–78%, $n = 5$); dorsal adductor scar quadripartite, 87% as long as wide (83–91%, $n = 3$) and 36% as long as the valve length (31–40%, $n = 3$); posterior adductor scars larger than the anterior scars.

Morphological variation. – *Hesperorthis leinsterensis* n. sp. was investigated for allometric growth. Valve length and width (L_valve and W_valve, respectively), brachiophore length and lateral spread (L_br and W_br, respectively), ventral muscle scar length and width (L_msc and W_msc, respectively) and ventral interarea length (L_int) included a sufficient amount of measurements for statistical analysis.

Valve dimensions

Total assemblage – valve width relative to valve length. – The log-transformed RMA plot (Fig. 23) of the valve widths plotted against valve length indicates no breakpoints; however, the four longest valves are separated from the rest of the assemblage by a gap in valve lengths between 10.4 and 12.7 mm (~1.01–1.10 on the log-transformed plot). Four specimens are not enough to divide the data into two size-groups and compare their individual regression slopes statistically and as these data points seem to be scattered around the regression line in a manner similar to the data points of the rest of the specimens they are analysed together as one assemblage. The log-transformed RMA indicates statistically significant allometry as $P(a = 1) < 0.05$. The length–width growth pattern is visualized by plotting the valve lengths against their corresponding valve length: width ratios ($L\_valve \times 100/W\_valve$) in a non-log-transformed RMA plot (Fig. 24). The non-transformed RMA plot indicates that the outline shape of *H. leinsterensis* n. sp. changed through ontogeny from small specimens with transversely wider than long valves, over equidimensional

larger specimens, to large mature individuals with valves longer than wide; that is, the valve length increased at a faster rate than the valve width.

Brachiophore dimensions

Brachiophore length relative to valve length. – The amount of measurements ($n = 10$) is insufficient to indicate breakpoints. The log-transformed RMA of the brachiophore lengths plotted against valve length (Fig. 25) produced contradictory outcomes when the x and y-coordinates were switched around and, hence, the corresponding ratio was tested against ontogenetic stage (the valve length) using the non-transformed RMA as the variables were not independent. The results of the RMA performed on the brachiophore length:valve length ratio ($L\_br \times 100/L\_valve$) against valve length are provided in Figure 26. This analysis is unfortunately also inconclusive as the data points are too scattered to indicate any trends.

Brachiophore length:lateral spread ratio relative to valve length. – The amount of measurements ($n = 10$) is insufficient to indicate breakpoints. The log-transformed RMA of the brachiophore length: lateral spread ratios plotted against valve length indicates that this ratio changed allometrically relative to valve length through ontogeny as $P(a = 1) < 0.05$

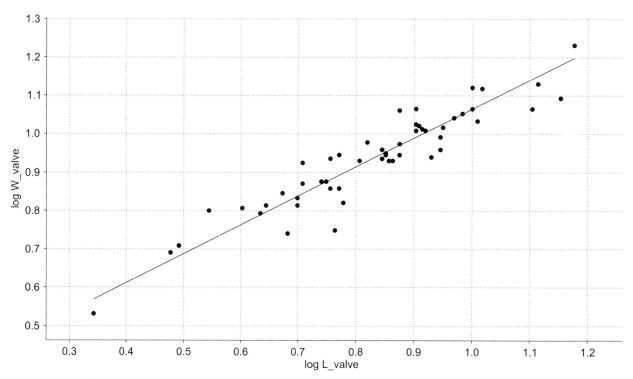

Fig. 23. *Hesperorthis leinsterensis* n. sp. Log-transformed RMA plot of the valve width plotted against the valve length. $N = 53$, slope $a = 0.75$, error $a = 0.04$, $r = 0.94$, $P(\text{uncorr}) = 4 \times 10^{-25}$, $P(a = 1) = 2 \times 10^{-8}$, significance level = 0.05. The log-transformed RMA indicates statistically significant allometry as $P(a = 1) < 0.05$.

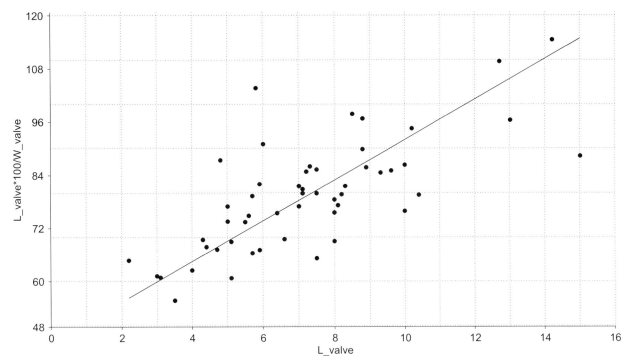

Fig. 24. Hesperorthis leinsterensis n. sp. RMA plot of the valve length:width ratio, i.e. the outline shape, (*L*_valve × 100/*W*_valve) plotted against the valve length. *N* = 53, slope *a* = 4.57, error *a* = 0.44, *r* = 0.73, *P*(uncorr) = 5 × 10⁻¹⁰, significance level = 0.05. The outline shape changed through ontogeny from immature individuals with greater valve width than length, over equidimensional larger specimens, to large, mature individuals with greater valve length than width.

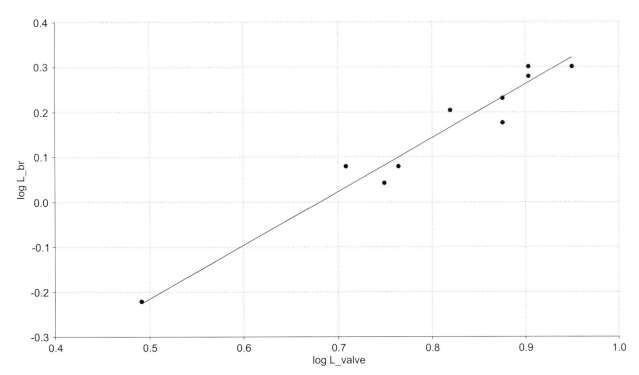

Fig. 25. Hesperorthis leinsterensis n. sp. Log-transformed RMA plot of the brachiophore length plotted against the valve length. *n* = 10, slope *a* = 1.19, error *a* = 0.09, *r* = 0.98, *P*(uncorr) = 1×10⁻⁶, *P*(*a* = 1) = 0.06, significance level = 0.05. Isometry cannot be rejected as *P*(*a* = 1) > 0.05 and the slope (*a*) is very close to 1. When the variables change places so the dorsal valve length is on the *y*-axis and the brachiophore length is on the *x*-axis a result of *P*(*a* = 1) = 0.03 is produced indicating allometry.

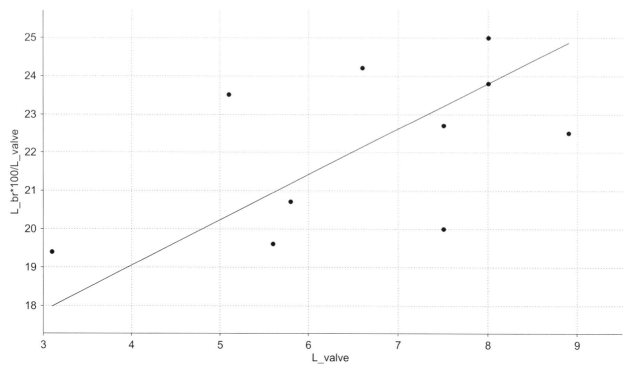

Fig. 26. Hesperorthis leinsterensis n. sp. RMA plot of the brachiophore length:valve length ratio ($L\_br \times 100/L\_valve$) plotted against the valve length. $n = 10$, slope $a = 1.19$, error $a = 0.36$, $r = 0.53$, P(uncorr) $= 0.12$, significance level $= 0.05$. The data points are very scattered and the result is inconclusive.

(Fig. 27). The brachiophore growth pattern is visualized in a non-log-transformed RMA plot of these variables (Fig. 28). The length of the brachiophores evolved from almost as long as their lateral spread in the smallest, immature specimens to about 65% as long as their lateral spread in the largest of the dorsal valves.

Ventral muscle scar dimensions
Muscle scar length relative to valve length. – The log-transformed RMA plot (Fig. 29) of the ventral muscle scar lengths plotted against valve length indicates no breakpoints. The log-transformed RMA shows that growth of the ventral muscle scar length relative to valve length was allometric as $P(a = 1) < 0.05$ (Fig. 29). The growth pattern of the muscle scar length is visualized in a non-log-transformed RMA plot of the ventral muscle scar length:valve length ratio ($L\_msc \times 100/L\_valve$) plotted against the valve length (Fig. 30). The regression line of the non-transformed RMA shows that the ratio increased through ontogeny indicating the growth rate of the ventral muscle scar length increased relative to valve length.

Muscle scar length:width ratio relative to valve length. – The log-transformed RMA plot (Fig. 31) of the ventral muscle scar width:length ratios

($W\_msc \times 100/L\_msc$) plotted against valve length indicates no breakpoints. The log-transformed RMA shows that the increase in ventral muscle scar length: width ratio relative to valve length was allometric as $P(a = 1) < 0.05$ (Fig. 31). The muscle scar growth pattern is visualized in the non-log-transformed RMA plot of these variables (Fig. 32), which clearly shows that the length of the muscle scar increases considerably relative to width through ontogeny. This changed the outline shape of the ventral muscle scar from wider than long in the smallest, immature individuals, over equidimensional scars in small specimens, to scars with much greater length than width in the largest, mature individuals.

Ventral interarea
Interarea length relative to valve length. – The log-transformed RMA plot (Fig. 33) of the ventral inter-area lengths plotted against valve length indicates no breakpoints. The result of the log-transformed RMA shows that growth of the ventral interarea relative to valve length was allometric as $P(a = 1) < 0.05$. The growth pattern of the ventral interarea is visualized in a non-log-transformed RMA plot of the ventral interarea length:valve length ratio ($L\_int \times 100/L\_$-valve) plotted against ventral valve length (Fig. 34). The non-transformed RMA indicates that the length of the ventral interarea increased relative to valve

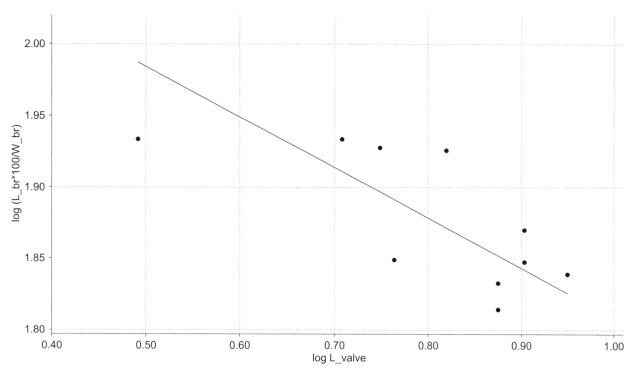

Fig. 27. *Hesperorthis leinsterensis* n. sp. Log-transformed RMA plot of the brachiophore length:width ratio ($L\_br \times 100/W\_br$) plotted against the valve length. $n = 10$, slope $a = -4.80$, error $a = 1.14$, $r = -0.74$, $P(\mathrm{uncorr}) = 0.01$, $P(a = 1) = 9 \times 10^{-4}$, significance level = 0.05. The log-transformed RMA indicates statistically significant allometry as $P(a = 1) < 0.05$.

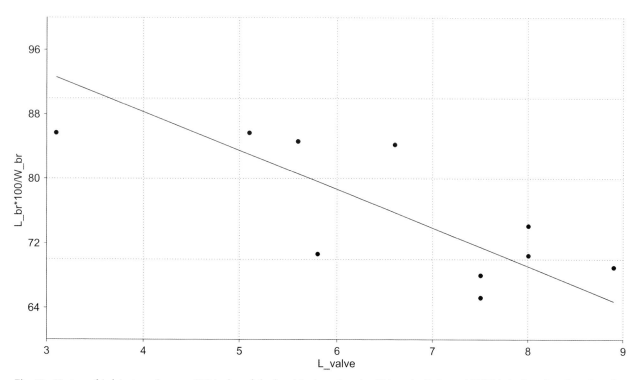

Fig. 28. *Hesperorthis leinsterensis* n. sp. RMA plot of the brachiophore length:width ratio ($L\_br \times 100/W\_br$) plotted against the valve length. $n = 10$, slope $a = -4.80$, error $a = 1.14$, $r = -0.74$, $P(\mathrm{uncorr}) = 0.01$, significance level = 0.05. The plot indicates that the lateral spread of the brachiophores increased relative to brachiophore length through ontogeny.

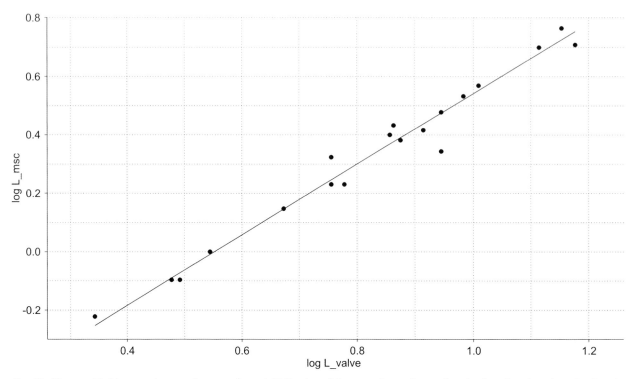

Fig. 29. Hesperorthis leinsterensis n. sp. Log-transformed RMA plot of the ventral muscle scar length plotted against the valve length. $n = 19$, slope $a = 1.21$, error $a = 0.05$, $r = 0.99$, $P(\text{uncorr}) = 4 \times 10^{-15}$, $P(a = 1) = 3 \times 10^{-4}$, significance level $= 0.05$. The log-transformed RMA indicates statistically significant allometry as $P(a = 1) < 0.05$.

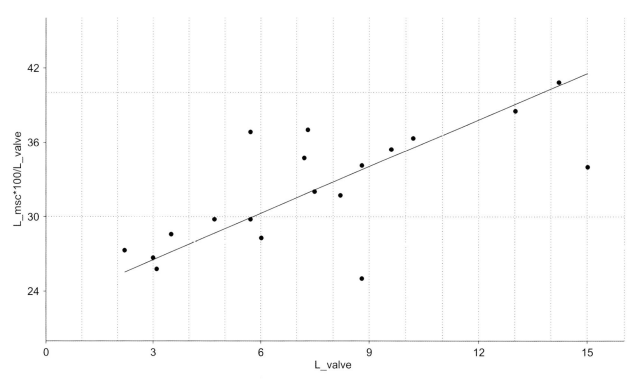

Fig. 30. Hesperorthis leinsterensis n. sp. RMA plot of the ventral muscle scar length:valve length ratio ($L\_msc \times 100/L\_valve$) plotted against the valve length. $n = 19$, slope $a = 1.25$, error $a = 0.22$, $r = 0.70$, $P(\text{uncorr}) = 8 \times 10^{-4}$, significance level $= 0.05$. The plot indicates that the ventral muscle scar increased in length relative to valve length through ontogeny.

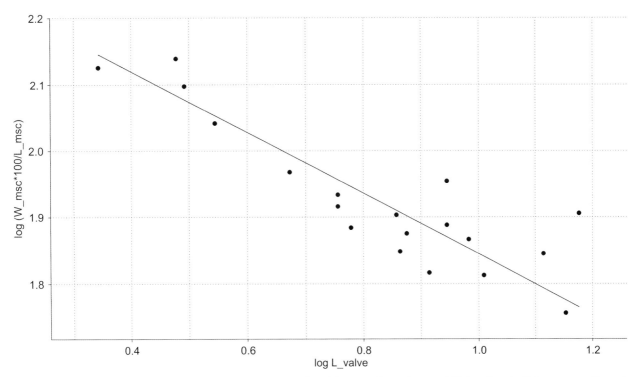

Fig. 31. Hesperorthis leinsterensis n. sp. Log-transformed RMA plot of the ventral muscle scar width:length ratio (W_msc $\times$ 100/L_msc) plotted against the valve length. $n = 19$, slope $a = -0.45$, error $a = 0.05$, $r = 0.70$, P(uncorr) $= 9 \times 10^{-7}$, $P(a = 1) = 2 \times 10^{-15}$ significance level $= 0.05$. The log-transformed RMA indicates statistically significant allometry as $P(a = 1) < 0.05$.

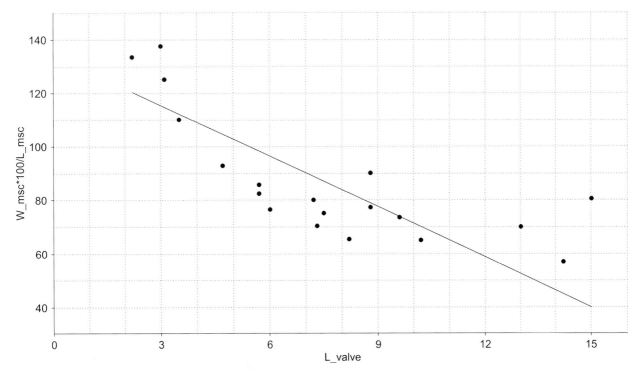

Fig. 32. Hesperorthis leinsterensis n. sp. RMA plot of the ventral muscle scar width:length ratio (W_msc $\times$ 100/L_msc) plotted against the valve length. $n = 19$, slope $a = -6.29$, error $a = 1.00$, $r = -0.75$, P(uncorr) $= 2 \times 10^{-4}$, significance level $= 0.05$. The plot indicates that the ventral muscle scar increased in length relative to width through ontogeny.

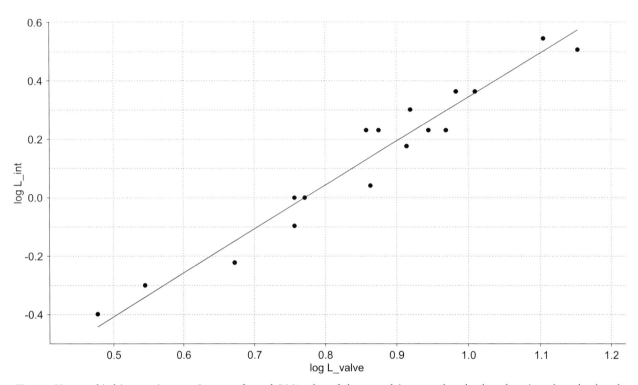

Fig. 33. Hesperorthis leinsterensis n. sp. Log-transformed RMA plot of the ventral interarea length plotted against the valve length. $n = 17$, slope $a = 1.50$, error $a = 0.09$, $r = 0.97$, $P(\text{uncorr}) = 5 \times 10^{-11}$, $P(a = 1) = 4 \times 10^{-05}$ significance level = 0.05. The log-transformed RMA indicates statistically significant allometry as $P(a = 1) < 0.05$.

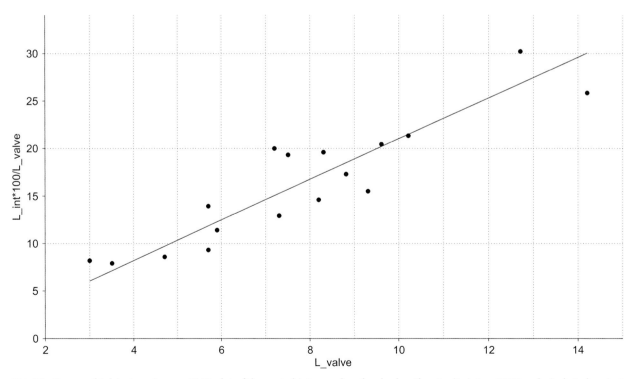

Fig. 34. Hesperorthis leinsterensis n. sp. RMA plot of the ventral interarea length:valve length ratio ($L\_int \times 100/L\_valve$) plotted against the valve length. $n = 17$, slope $a = 2.14$, error $a = 0.23$, $r = 0.91$, $P(\text{uncorr}) = 5 \times 10^{-7}$, significance level = 0.05. The plot indicates that the length of the ventral interarea increased relative to valve length through ontogeny.

length through ontogeny as the interarea length:-valve length ratio increases with greater valve length.

Summary

In summary, the valve outline shape as well as the dorsal brachiophores, ventral muscle scar and ventral interarea experienced allometric growth through ontogeny. The valve outline changed from wider than long to longer than wide; the lateral spread of the brachiophores increased; the ventral muscle scar and interarea length increased relative to valve length; and the ventral muscle scar furthermore changed outline shape from wider than long to longer than wide. The analysis of the brachiophore length relative to valve length was inconclusive.

Discussion. – In species of *Hesperorthis*, the number of ribs present is commonly diagnostic. Williams (1962) and Candela (2003) both commented on the great variation in the number of costae within the genus. Rib counts for species most similar to the Tramore form are provided in Table 12. Rib counts should not be the sole character used for species identification, however, as exemplified by the relation between *H. australis* Cooper, 1956, *H. australis exitis* Williams, 1962 and *H. craigensis* (Reed, 1917). *Hesperorthis craigensis* and *H. australis* have similar rib counts (Table 12), but as noted by Williams (1962) these species are morphologically more different than *H. australis* and its subspecies, which have themselves different rib counts.

Hesperorthis leinsterensis n. sp. is most similar to *H. australis exitis* and *H. craigensis*: *H. australis exitis* has been described from the Confinis Flags, uppermost Darriwilian (upper *TS*.4c), Bougang, Girvan, Scotland (Williams 1962) and from the Bardahessiagh Formation, Member I and III, upper Sandbian – lowermost Katian (middle *TS*.5b–middle *TS*.5c), Pomeroy, North Ireland (Candela 2003). It is very similar to *H. leinsterensis* n. sp. but can be distinguished by its finer ribbing with commonly 30–33 primary costae (Table 12) and its dorsal sulcus, which is distinct all the way to the anterior margin.

Furthermore, the broad dorsal median ridge disappears about mid-valve length in *H. australis exitis*, whereas it extends anteriorly for about 75% of the valve length in *H. leinsterensis* n. sp. Candela (2003) described two less elevated ridges flanking the cardinal process in *H. australis exitis*, which were referred to as 'lateral bosses' by Cooper (1956) in his description of *H. australis*. These were not observed in our specimens and Candela commented that these features have not yet been described for any other species of *Hesperorthis*.

Hesperorthis craigensis was redescribed by Williams (1962) based on specimens from the Craighead Limestone Formation, middle Sandbian – lower Katian (lowermost *TS*.5b–upper *TS*.5c), Craighead Quarry, Girvan, Scotland. This species is also very similar to *H. leinsterensis* n. sp. but, as *H. australis exitis*, it is more finely costate commonly having 26–30 primary costae. It can furthermore be distinguished on the shape of the ventral muscle scar being subcordate and about as wide as long for a sample including both young and adult specimens ($n = 23$) compared to the elongately triangular muscle scar of *H. leinsterensis* n. sp. with an average muscle scar width:length ratio of 87% (57–138%, $n = 19$).

Hesperorthis dynevorensis Williams, 1949; was redescribed by Lockley & Williams (1981) based on material from the upper part of the Pebbly Sands Formation of the Ffairfach Group, middle Darriwilian (*TS*.4b–middle *TS*.4c), Ffairfach, Llandeilo, Wales. This species is rather similar to *H. leinsterensis* n. sp. except for its finer ribbing with up to 35 primary costae (Table 12) together with its ventral adductor scar projecting further anteriorly than the diductor scars and the dorsal adductor scar being longer than wide with a length of 60% of the valve length compared to 36% (31–40%, $n = 3$) for *H. leinsterensis* n. sp.

Hesperorthis inostrantzefi (Wysogórski, 1900) from the Kukruse Formation, lower Sandbian (*TS*.5a), was revised by Öpik (1930) from C2 to C3 beds, Kohtla, Vanamõisa and Sala, Estonia, and later

Table 12. Rib counts for *Hesperorthis leinsterensis* n. sp. and resembling species.

Species	Data source	20	22	24	25	26	27	28	29	30	31	32	33	34	35
H. australis (Girvan)	Candela (2003)	–	–	–	–	1	3	[9]	**4**	3	3	1	0	–	–
H. australis exitis (Girvan)	Candela (2003)	–	–	–	–	–	–	1	3	[9]	**3**	3	3	–	–
H. australis exitis (Pomeroy)	Candela (2003)	–	–	–	–	–	–	–	–	[7]	**2**	2	4	–	1
H. craigensis (Girvan)	Candela (2003)	–	–	–	1	5	6	[9]	**6**	7	4	3	2	–	–
H. dynevorensis (Builth)	Lockley & Williams (1981)	–	–	–	2	3	6	[7]	**6**	2	1	–	–	–	–
H. dynevorensis (Llandeilo)	Lockley & Williams (1981)	–	–	–	1	–	–	1	3	4	[6]	**4**	1	–	2
H. inostrantzefi (Estonia)	Öpik (1930)	1	1	**1**	[3]	1	–	1	–	–	–	–	–	–	–
H. leinsterensis n. sp. (Tramore area)	This study	–	–	4	[9]	–	–	–	–	–	–	–	–	–	–

The mean is shown in bold, the mode in brackets.

described by Harper (1952) from Caradoc rocks at Grangegeeth, Co. Meath, east Ireland (Cocks 2008). *Hesperorthis inostrantzefi* from Estonia (Öpik 1930) and *H. leinsterensis* n. sp. have similar rib counts of 24–25 costae (Table 12) but they are otherwise very different. The most pronounced differences shown by the Estonian form is the almost catacline, curved interarea on adult ventral valves and the torpedo-shaped ventral muscle scar with the widest part in the posterior half of the scar. Furthermore, the primary costae are flattened and quadrate in cross-section. *Hesperorthis inostrantzefi* from east Ireland has 30 primary costae and is more finely costate than the Tramore form.

Cooper (1956) formally described 16 species of *Hesperorthis* but none of these are similar to *H. leinsterensis* n. sp. The most similar North American species is *H. australis* from the Dryden, Wardell and Ridley formations, upper Sandbian – lower Katian (base of *TS*.5c–upper middle *TS*.5c), Appalachians, eastern USA. This species was described as the North American species most closely related to the British species by Williams (1949, 1962) and Candela (2003) which is corroborated by this study. It differs from *H. leinsterensis* n. sp. in commonly having 26–30 costae, a concavoconvex profile, unequally convex ventral valve, straight lateral margins, distinct dorsal sulcus, long brachiophores, a dorsal median ridge extending to about mid length, and a stout cardinal process flanked by 'lateral bosses'.

Parkes (1994) recorded three ventral valves of *Hesperorthis* sp. from the Duncannon Group, uppermost Sandbian (lower *TS*.5c), Kilbride, Co. Waterford, southeast Ireland. He commented that the species has a long apsacline interarea and is smaller than, and morphologically distinct from, *H. inostrantzefi*. The ventral valve illustrated by Parkes (1994, pl. 3, fig. 10) is about 4.4 mm long and 4.4 mm wide with approximately 22 rounded primary costae, although these are obscured due to poor preservation. It furthermore has a ventral muscle scar about as long as wide possibly with the adductors extending anteriorly of the diductors, but these details are likewise obscured. The equidimensional valves and ventral muscle scar as well as the valve size is within the morphological range of *H. leinsterensis* n. sp. but if the *Hesperorthis* sp. described by Parkes (1994) only has 22 primary costae and adductors extending anteriorly of the diductors the two forms are not conspecific. More material of *Hesperorthis* sp. is needed to clarify this.

Occurrence. – Tramore Limestone Formation: Units 1–3 and 5, Barrel Strand area, Pickardstown section and Quillia Section 2. Dunabrattin Limestone Formation, Unit II, Dunabrattin–Bunmahon area).

Family Plaesiomyidae Schuchert, 1913
Subfamily Plaesiomyinae Schuchert, 1913

Genus *Valcourea* Raymond, 1911

Type species (by original designation). – *Plaesiomys strophomenoides* Raymond, 1905, from the Crown Point Formation, Sandbian–lower Katian (*TS*.5a–upper *TS*.5d), New York, USA.

Valcourea confinis (Salter, 1849)

Plate 6, figures 14–17, 19–20

1849   *Orthis confinis* Salter, p. 15, pl. 1, fig. 4.

1869   *Orthis confinis* Salter; Davidson, *pars*, pl. 36, figs 1–3, *non* fig. 4.

1871   *Orthis confinis* Salter; Davidson, p. 266 *pars*.

1917   *Orthis (Heterorthis?) confinis* Salter; Reed, p. 847, pl. 8, figs 26–30, 32, 33.

1962   *Valcourea confinis* (Salter); Williams, p. 117, pl. 10, figs 8, 10, 11, 13, 14.

1978   *Valcourea confinis* (Salter); Cocks, p. 52.

2008   *Valcourea confinis* (Salter); Cocks, p. 131.

Comments on the synonymy list. – The specimen figured by Davidson (1869, pl. 36, fig. 4) probably belongs to a different species as it does not have the characteristic bulbous cardinal process of *V. confinis* but the drawing is not sufficiently informative and key details are obscured.

Material. – Total of three ventral valves and three dorsal valves.

Description. – Suboval, convexo-concave to convexo-planar *Valcourea* species, 5–15 mm long ($n = 5$) and 9–21 mm wide ($n = 5$) with maximum width along the hingeline; cardinal angles just less than 90°; lateral and anterior margins gently and evenly rounded; narrow ventral fold and complementary dorsal sulcus arise sharply at the umbones but become less pronounced anteriorly and die out in adult forms. Ventral valve planar to gently concave, 59% as long as wide (59–

60%, *n* = 2); dorsal valve gently convex with maximum depth posterior of the valve centre; 69% as long as wide (64–73%, *n* = 3), 19% as deep as long (18–20%, *n* = 2) and 13% as deep as wide (range = 0%, *n* = 2); ventral interarea long, apsacline with pseudodeltidium weakly developed apically and laterally; dorsal interarea short, anacline with arched chilidium apically; ornament finely multicostellate with aditicules; second and third order costellae developed in adults; counts of 4 costellae per mm at 5 mm anterior to the umbo for two dorsal and two ventral valves, respectively, and 17 and 19 costellae per 5 mm at 5 mm anterior to the umbo for three dorsal and one ventral valve, respectively; radial ornamentation cancellated by closely spaced, delicate filae.

Ventral interior represented by only one specimen distorted by stretching across its width; teeth short, peglike; ventral muscle scar bilobed in outline, which is typical of *Valcourea*, but with distorted dimensions; diductor scars divergent, pointed posteriorly and widening anteriorly; adductor scar elongately triangular; subperipheral rim well-developed; mantle canals not impressed.

Dorsal interior with cardinal process consisting of a shaft and a strong, bulbous, weakly trilobed myophore with a slightly carinate median lobe; sockets large, well-developed and laterally flaring located posteriorly of the brachiophores; brachiophores rodlike and diverging at about 90°, joined to the valve floor for most of their length, 60% as long as their lateral spread (53–66%, *n* = 2) and 17% as long as valve length (14–19%, *n* = 2); median septum continuous with the shaft of the cardinal process and extending forwards for 40% of the valve length (36–43%, *n* = 2) to bisect the adductor muscle scars, which are weakly impressed posteriorly but obscured anteriorly and laterally; mantle canals not impressed.

Discussion. – This is the first time *Valcourea* has been described from southeast Ireland. The material from Tramore is assigned to *V. confinis* (Salter, 1849) which was described in detail by Williams (1962, p. 117) from the basal Balclatchie Conglomerate Formation, middle to upper Sandbian (lowermost *TS.5b*–middle *TS.5c*), Brockloch, Girvan, Scotland. Williams (1962) provided ratios used in his taxonomical description of the Girvan species, which are compared to the Tramore form together with rib counts measured on Williams' specimens (pl. 10, figs 8, 14) (see Tables 13 and 14, respectively). The two forms are of similar size with almost identical ratio means and identical rib counts. They are also similar in their suboval

outline, short pseudodeltidium, cardinal process and sockets and the sharp ventral fold and dorsal sulcus terminating posteriorly of the anterior margin in adults.

Valcourea confinis is closely comparable to *Valcourea semicarinata* Cooper, 1956, described from the upper part of the Arline Formation, uppermost Darriwilian (upper *TS.4c*), Porterfield Quarry, Virginia. Williams (1962, p. 118) commented that *V. semicarinata* might be conspecific with *V. confinis s.s.*; *Valcourea semicarinata* has a costellate ornament similar to the Tramore form with 3–4 costellae per mm at the anterior margin. It also has a short pseudodeltidium, a shallow dorsal valve with sulcus extending about 50% of the valve length, and a similar shaped ventral adductor and diductor scars. It can be distinguished from the Tramore and Girvan forms by its significantly longer dorsal median ridge extending for about 75% the valve length.

Occurrence. – Tramore Limestone Formation, Units 2–3, Barrel Strand area.

Family Productorthidae Schuchert & Cooper, 1931

Subfamily Productorthinae Schuchert & Cooper, 1931

Genus *Productorthis* Kozłowski, 1927

Type species (by original designation). – *Productus obtusus* Pander, 1830, from the Volkhov Formation, Dapingian–lower Darriwilian (uppermost *TS.3b*–*TS.4a*), St. Petersburg Region, Russia.

Productorthis sp.

Plate 6, figure 21

Material. – One dorsal valve.

Remarks on the material. – The specimen at hand consisted of a dorsal valve with partly preserved exteriors covering an interior mould. The interior mould was prepared by dissolving away the exteriors destroying all traces of these in the process. The specimen appears to be slightly distorted by postmortem deformation but it is unknown how much this has affected the valve dimensions.

Description. – The available dorsal valve is transversely subcircular with obtuse cardinal angles and rounded lateral and anterior margins; 13 mm long, 18 mm wide and 74% as long as wide with maximum width occurring medially; in profile, the umbonal area appears almost flat but the valve is otherwise strongly convex, 40% as deep as long with maximum convexity occurring about medially; ornament costellate with strong concentric lamellae developed anteriorly.

Dorsal interior with a massive cardinal process filling the notothyrium and buttressed by a pair of slightly divergent, short brachiophores; dental sockets deep, crescent-shaped and strongly transverse, subparallel to the brachiophores and curving inwards towards the cardinal process; adductor muscle scar large, quadripartite, as long as wide and 53% as long as the valve length, separated medially by a thin median ridge; posterior adductor scars small and transversely rectangular; anterior adductor scars large and lobate with the larger lobe located medially.

Discussion. – The Tramore specimen is reminiscent of the dorsal valves of *Productorthis mitchelli* Williams, 1956, from the Upper Tuffs and Shales Formation of the Grangegeeth Volcanic Series middle–upper Sandbian (lowermost *TS*.5b–middle *TS*.5c), Grangegeeth, Co. Meath, east Ireland. Their dorsal adductor scars are thus both equidimensional and similar in shape; however, the scar of *P. mitchelli* is only 39% as long as the valve length measured on Williams figured specimen (1956, pl. 9, fig. 4). The outline of the Tramore form appears to be more transverse and more strongly convex than the Grangegeeth form, which is 87% as long as wide (84–90, $n = 2$) and 26% as deep as long ($n = 1$) measured on Williams' figured specimens (1956, pl. 9, figs 3, 4; fig. 1). The slightly distorted nature of the Tramore specimen makes comparison of the valve dimensions uncertain, however. The Tramore specimen furthermore differs from *P. mitchelli* in that its cardinal process fills the notothyrium but this difference may have been caused by excessive shell secretion.

Another, probably closely related, British species, *Productorthis lamellosa* (Bates, 1968) was described from the Treiorwerth Formation, Dapingian–lower Darriwilian (*TS*.3a–*TS*.4a), Ffynnon-y-mab, Trefor, Anglesey, North Wales. Neuman & Bates (1978) noted some slight differences between specimens from the Treiorwerth and Bod Deiniol formations, the Bod Deiniol valves being the larger ones with a more pronounced dorsal median ridge and more deeply impressed dorsal adductor scars. The Tramore specimen differs from the dorsal valve figured by Bates (1968, pl. 3, figs 11, 12) in having a much thinner median ridge and shorter and less divergent brachiophores. It differs from the dorsal valves illustrated by Neuman and Bates (1978, pl. 68, figs 1, 2; pl. 63, figs 5, 6) in having shorter and less divergent brachiophores but they have similar median ridges.

Occurrence. – Tramore Limestone Formation, Quillia Section 2.

Table 13. Comparison of selected ratios of *Valcourea confinis* from Tramore and Girvan.

Ratio	Valve	Sample	Data source	n	Mean (%)	Range (%)
L_valve × 100/W_valve	Ventral	Tramore	This study	2	59	59–60
		Girvan	Williams (1962)	34	61	–
W_msc × 100/L_msc	Ventral	Tramore	This study	1	(146)	–
		Girvan	Williams (1962)	10	90	–
L_msc × 100/L_valve	Ventral	Tramore	This study	1	40	–
		Girvan	Williams (1962)	8	44	–
D_valve × 100/W_valve	Dorsal	Tramore	This study	2	13	–
		Girvan	Williams (1962)	12	15	–
L_ms × 100/L_valve	Dorsal	Tramore	This study	2	40	36–43
		Girvan	Williams (1962)	–	<40	–

L, length; W, width; msc, muscle scar; ms, median septum.

Table 14. Comparison of rib counts of *Valcourea confinis* from Tramore and Girvan.

Character	Valve	Sample	Data source	n	Mean	Range
No. of costellae per mm at 5 mm anterior to umbo	Ventral + dorsal	Tramore	This study	4	4	–
		Girvan	Williams (1962)	2	4	3–4
No. of costellae per 5 mm at 5 mm anterior to umbo	Ventral + dorsal	Tramore	This study	4	17	17–19
		Girvan	Williams (1962)	2	17	–

The number of ribs of the Girvan form was counted on the specimens of Williams (1962, pl. 10, figs 8, 14).

Superfamily Plectorthoidea Schuchert & LeVene, 1929

Family Platystrophiidae Schuchert & LeVene, 1929, emend. Zuykov & Harper, 2007

Genus *Platystrophia* King, 1850, emend. Zuykov & Harper, 2007

Type species (by subsequent designation). – *Porambonites costata* Pander, 1830; from the Obukhovo Formation, upper Sandbian (*TS*.5a), St Petersburg Region, northwest Russia. Designated by Zuykov & Harper 2004; discussed by Zuykov & Harper 2007.

Discussion. – Zuykov & Harper (2007) revised the genus *Platystrophia* and restricted it to a group of Arenig to upper Caradoc species from Baltica and Avalonia, whereas the Ashgill and Silurian taxa from these regions, previously assigned to *Platystrophia*, were placed in the new genus *Neoplatystrophia*.

Platystrophia tramorensis n. sp.

Plate 6, figure 18; Plate 7, figures 1–9, 11; Plate 10, figure 8

1994 *Platystrophia* sp. 2 Parkes, p. 143, pl. 5, figs 4, 5.

Name. – Alluding to the Tramore Limestone Formation.

Holotype. – NHMUK PI BB35271, interior mould of dorsal valve, Tramore Limestone Formation (Pl. 7, figs 1, 2).

Material. – Total of 28 ventral valves and 27 dorsal valves.

Diagnosis. – Transversely subquadrate to subcircular *Platystrophia* species; profile dorsibiconvex and rather flattened; valves 68% as long as posteromedially occurring maximum width, ventral and dorsal valves 33% and 45% as deep as long, respectively; ventral sulcus and dorsal fold 21% as wide as maximum valve width at 5 mm anterior to the umbo; radial ornamentation with two strong costae in the sulcus, three on the fold and 7–10 on the flanks with a mode of 8; ventral muscle scar 50% as wide as long and 40% as long as the valve length; cardinal process

a simple ridge; brachiophores stout and peglike, dorsal adductor scar 87% as long as wide and 37% as long as the valve length bisected by a strong median ridge and a pair of widely divergent transverse ridges.

Description. – Transversely subquadrate to subcircular *Platystrophia* species with bluntly rectangular cardinal angles; profile dorsibiconvex and rather flattened; anterior profile with uniplicate commissure; 4–12 mm long ($n = 50$) and 7–18 mm wide ($n = 51$) with maximum valve width occurring posteromedially; hingeline 86% as wide as maximum width (73–99%, $n = 37$); conjoined valves of equal length and width, 68% as long as maximum width (43–92%, $n = 48$); ventral valve 33% as deep as long (22–44%, $n = 14$) and 24% as deep as wide (14–38%, $n = 16$); dorsal valve 45% as deep as long (31–63%, $n = 10$) and 31% as deep as wide (23–37%, $n = 14$); ventral and dorsal interareas short, apsacline and orthocline, respectively; delthyrium and notothyrium open; well-defined ventral sulcus and dorsal fold of equal width in conjoined valves, 21% as wide as the maximum valve width at 5 mm anterior to the umbo (13–31%, $n = 39$); ornament consisting of fine, closely spaced spine bases and strong, sharply crested, radiating costae with a total of two costae in the sulcus ($n = 21$), three on the fold ($n = 24$) and 7–11 on each flank, but commonly with eight ($n = 22$ of 56 flanks) or nine ($n = 17$ of 56 flanks) costae on each flank, although a specimen may have one more costa on one flank compared to the other.

Ventral interior with large prominent teeth and crural fossettes supported by slightly divergent dental plates, which are continuous with a U-shaped ridge defining the anterolateral and anterior limits of the ventral muscle scar; muscle scar 50% as wide as long (34–70%, $n = 12$) and 40% as long as the valve length (31–52%, $n = 12$).

Dorsal interior (Pl. 10, fig. 8) with a simple, ridgelike cardinal process extending the entire length of the small notothyrial platform to meet with a strong median ridge; dorsal sockets medium sized and distinct; brachiophores stout, peglike and divergent; adductor scar quadripartite, subcircular and deeply impressed, 87% as long as wide (66–111%, $n = 8$) and 37% as long as the valve length (28–53, $n = 7$), bisected by the strong median ridge and a pair of widely divergent transverse ridges, the median ridge extending to the anterior end of the adductor scar; posterior adductor scars rectangular to subtriangular with their long axis oriented laterally, smaller than the anterior scars; anterior adductor scars triangular and lobate, widest at their anterolateral margin with the larger lobe located medially.

Morphological variation. – *Platystrophia tramorensis* n. sp. was investigated for allometric growth. Valve length and maximum width (L_valve and W_valve, respectively), hinge width (W_hi), ventral and dorsal valve depth (D_valve), ventral and dorsal muscle scar length and width (L_msc and W_msc, respectively), and sulcus and fold width at 5 mm anterior to the umbo (W_sf), included a sufficient amount of measurements for morphometric analysis. The results of the log-transformed RMAs performed on these variables are provided in Table 15. No breakpoints were observed and $P(a = 1) > 0.05$ in all the analyses indicating that the species maintained an isometric growth trajectory for all the tested variables.

The log-transformed RMAs performed on the valve widths and dorsal valve depths plotted against valve length gave contradictory outcomes when the x and y-coordinates were switched around and, hence, their corresponding ratios were tested against ontogenetic stage (the valve length) using the non-transformed RMA as the variables were not independent. The non-transformed RMA plots of the valve length:width ratio and dorsal valve depth:length ratio plotted against valve length are shown in Figures 35 and 36, respectively, and the statistical results are listed in Table 15. The RMAs support that the growth of these variables was isometric as the data points are randomly scattered around the regression lines and no changes in distributional patterns of the data points, that is possible breakpoints, are observed.

Discussion. – Two species of *Platystrophia* have been described from the Tramore area but they do not occur together and appear to have been confined to different palaeoenvironments. *Platystrophia tramorensis* n. sp. has only been observed in the shelf facies of the Tramore Limestone Formation and

Platystrophia aff. *P. sublimis* has only been recorded from the deep-water facies of the Dunabrattin Limestone Formation. Although they are rather similar in their valve ratios and the expression of their interiors, they are, with a few exceptions, readily distinguished based on their rib counts, as *P.* aff. *P. sublimis* has three costae in the sulcus, four on the fold and 6–9 on each flank with a mode of seven.

Platystrophia tramorensis n. sp. is not similar to any of the species assigned to *Platystrophia* by Zuykov & Harper (2007, p. 24) and by Cocks (2008), except for *Platystrophia* sp. 2 (see below). Parkes (1994) described *Platystrophia* sp. 2 from the Tramore Limestone Formation, Kilbride, Co. Waterford, southeast Ireland. It is similar to *P. tramorensis* n. sp. in both external ornament and dorsal interiors and was collected from the same formation and area as the current material. Parkes did not figure any ventral valves, but his form is very likely conspecific with *P. tramorensis* n. sp.

Of the non-Irish species, the Tramore form appears to be most similar to *Platystrophia dentata veimarnensis* Alikhova, 1951, described from the early Sandbian, Leningrad Area, northwest Russia. They have similar rib counts, the Russian form having two costae in the sulcus, three on the fold and 8–9 on the flanks as well as similar outline shape. *Platystrophia dentata veimarnensis* differs from *P. tramorensis* n. sp. in commonly being widest at the hingeline, and it has an almost imperceptible dorsal interarea. Alikhova (1951, p. 12) did not describe any interiors or valve sizes and, hence, it is not possible to compare these characters.

Platystrophia tramorensis n. sp. has the same number of costae in the sulcus and on the fold as the much discussed *Platystrophia dentata* (Pander, 1830) emend. Zuykov *et al.* (2011), but is otherwise very different from this species. Many species of *Platystrophia* now identified by Zuykov *et al.* (2011) as

Table 15. *Platystrophia tramorensis* n. sp. Results of log-transformed RMAs.

Characters and ratios	Valve	n	Slope a	Error a	r	P(uncorr)	$P(a = 1)$
W_valve∧L_valve	Ventral + dorsal	48	1.23	0.12	0.73	3×10^{-9}	0.07
W_hi∧W_valve	Ventral + dorsal	39	1.08	0.05	0.96	6×10^{-22}	0.13
D_valve∧L_valve	Ventral	16	1.24	0.26	0.61	0.01	0.37
	Dorsal	13	1.55	0.35	0.67	0.01	0.14
D_valve∧W_valve	Ventral	16	0.94	0.21	0.57	0.02	0.77
	Dorsal	14	1.24	0.22	0.79	9×10^{-4}	0.30
L_msc∧W_msc	Ventral	14	1.01	0.22	0.67	9×10^{-3}	0.95
	Dorsal	10	0.96	0.20	0.82	4×10^{-3}	0.85
L_msc∧L_valve	Ventral	14	1.14	0.18	0.84	2×10^{-4}	0.45
	Dorsal	10	1.45	0.38	0.68	0.03	0.26
W_sf∧W_valve	Ventral + dorsal	40	0.93	0.14	0.42	7×10^{-3}	0.62
L_valve × 100/W_valve	Ventral + dorsal	48	6.96	1.02	0.10	0.49	–
D_valve × 100/L_valve	Dorsal	10	6.85	2.41	0.09	0.80	–

∧, plotted against; L, length; W, width; hi, hingeline; msc, muscle scar; sf, sulcus and fold. Analyses, including ratios, were not log-transformed.

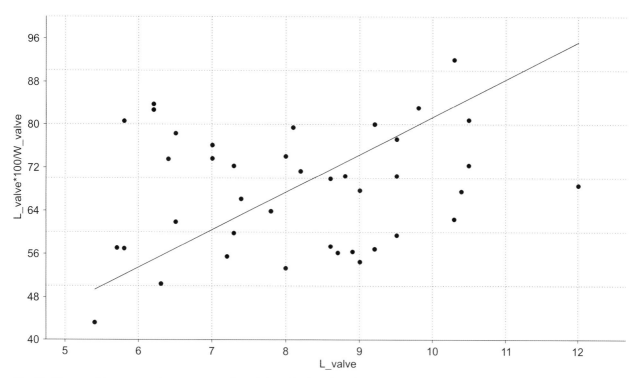

Fig. 35. Platystrophia tramorensis n. sp. RMA plot of the valve length:width ratio (*L*_valve × 100/*W*_valve) plotted against the valve length. *n* = 48, slope *a* = 6.96, error *a* = 1.02, *r* = 0.10, *P*(uncorr) = 0.49, significance level = 0.05.

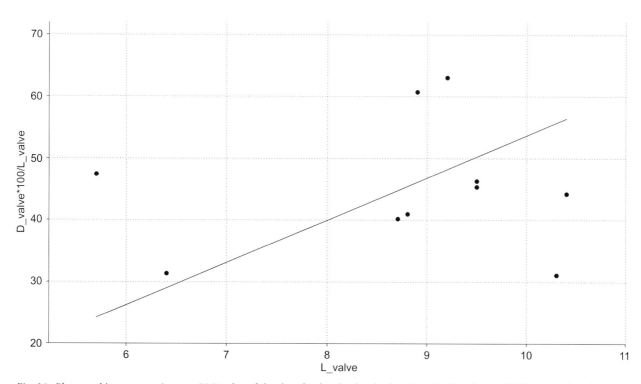

Fig. 36. Platystrophia tramorensis n. sp. RMA plot of the dorsal valve depth:valve length ratio (*D*_valve × 100/*L*_valve) plotted against the valve length. *n* = 10, slope *a* = 6.85, error *a* = 2.41, *r* = 0.09, *P*(uncorr) = 0.80, significance level = 0.05.

not belonging to *dentata s.s.* have previously been assigned to this species based on the number of costae in the sulcus and on the fold. *Platystrophia*

dentata was recollected and described by Zuykov *et al.* (2011) from the Simankovo and Duboviki formations, upper middle Darriwilian (lowermost

*TS.*4c), St. Petersburg Region, northwest Russia. Although both *P. tramorensis* n. sp. and *P. dentata* have two costae in the sulcus and three on the fold, the latter is more coarsely costellate with only 5–6 costae on each flank. *Platystrophia dentata* is also more strongly convex and the dorsal interiors are different with weakly impressed adductor scars, each of the four scars being elongately suboval to subquadrate and oriented subparallel to the median ridge. Furthermore, in *P. dentata*, only the anterior adductor scars are separated by a short, thin median ridge, and the posterior and anterior adductor pairs are not separated by a pair of transverse ridges.

Occurrence. – Tramore Limestone Formation, Units 2–3, Barrel Strand area.

Platystrophia aff. *P. sublimis* Öpik, 1930

Plate 7, figures 10, 12–18

aff. 1930 *Platystrophia sublimis* Öpik, p. 108, pl. 5, fig 50.

Material. – Total of 14 ventral valves and 10 dorsal valves.

Description. – Transversely subquadrate to subcircular *Platystrophia* species with bluntly rectangular cardinal angles; profile dorsibiconvex and rather flattened; 3–9 mm long ($n = 16$) and 6–13 mm wide ($n = 16$) with maximum valve width occurring posteromedially, one specimen was exceptionally wide being 19 mm but the valve length was obscured and not measurable; hingeline 94% as wide as maximum width (89–95%, $n = 6$); conjoined valves of equal length and width, 70% as long as wide (52–91%, $n = 15$); ventral valve 33% as deep as long (31–35%, $n = 2$) and 23% as deep as wide (20–27%, $n = 2$); dorsal valve 39% as deep as long (39–40%, $n = 2$) and 28% as deep as wide (27–30%, $n = 2$); ventral and dorsal interareas short, apsacline and orthocline respectively and with delthyrium and notothyrium open; well-defined ventral sulcus and dorsal fold of equal width in conjoined valves, 23% as wide as the maximum valve width (16–31%, $n = 7$) at 5 mm anterior to the umbo; ornament consisting of fine, closely spaced spine bases and strong, sharply crested, radiating costae with commonly three costae in the sulcus (2–3, $n = 11$) one of them arising by bifurcation from a main costa, four on the fold ($n = 10$) and 6–9

on the flanks, but commonly with seven costae on each flank ($n = 14$ of 29 flanks), although a specimen may have one more costa on one flank compared to the other.

Ventral interior with stout teeth and prominent crural fossettes supported by slightly divergent dental plates which are continuous with a U-shaped ridge defining the anterolateral and anterior margins of the ventral muscle scar; muscle scar 49% as wide as long (39–64%, $n = 3$) and 35% as long as the valve length (27–42%, $n = 3$).

Dorsal interior with a simple, ridgelike cardinal process extending the entire length of the small notothyrial platform to meet with a strong median ridge; dorsal sockets large and deeply impressed; brachiophores stout, peglike, divergent and strongly developed; adductor scar quadripartite, subcircular and deeply impressed, 89% as long as wide (88–89%, $n = 3$) and 40% as long as the valve length (23–58%, $n = 2$), bisected by the strong median ridge and a pair of widely divergent transverse ridges, the median ridge extending to the anterior end of the adductor scar; posterior adductor scars rectangular to almost quadratic with their long axis directed laterally, smaller than the anterior scars; anterior adductor scars triangular and widest at their anterolateral margin.

Discussion. – The Dunabrattin form is morphologically close to both *P. sublimis* Öpik, 1930, from Estonia and *Platystrophia* cf. *P. sublimis* Öpik, 1930, described from Wales by Williams (1963). Their valve dimensions and ratios are compared in Tables 16 and 17, respectively. *Platystrophia* aff. *P. sublimis* is most similar to *P.* cf. *P. sublimis* from the Gelli-Grîn Group, upper Sandbian (lowermost *TS.*5b–middle *TS.*5c), Bala, Gwynedd, Wales. They have similar rib counts as well as comparable ratios of valve length:width, ventral muscle scar width:length and ventral muscle scar length:valve length (Table 17). The Welsh form, however, is larger with valve lengths up to 27 mm, it has a more convex ventral valve being 50% as deep as the valve length, and almost equally dimensioned, subtriangular dorsal adductor scars separated by a much thicker median ridge and transverse ridges than observed in the Dunabrattin form. *Platystrophia sublimis* was originally described from the Shale Quarry, lower Sandbian (~*TS.*5a), Kohtla, Estonia. The Estonian and Dunabrattin forms have equal rib counts and appear to be of similar size, although the one pair of conjoined valves described and figured by Öpik (1930) is not enough to state this with certainty. *Platystrophia sublimis* can be distinguished

Table 16. Measurements of *P. sublimis* (Öpik 1930), *P.* cf. *P. sublimis* (Williams 1963), and *P.* aff. *P. sublimis* (this study).

Character	Valve	Species	Data source	n	Mean (mm)	Range (mm)
L_valve	Ventral + dorsal	*P. sublimis*	Öpik (1930)	1	7.0	–
		P. cf. *P. sublimis*	Williams (1963)	4	18.1	5.5–27.0
		P. aff. *P. sublimis*	This study	16	6.1	3.2–9.0
W_valve	Ventral + dorsal	*P. sublimis*	Öpik (1930)	1	12.0	–
		P. cf. *P. sublimis*	Williams (1963)	1	9.5 (L = 5.5 mm)	–
		P. aff. *P. sublimis*	This study	15	8.9	5.0–13.3
L_hi	Ventral + dorsal	*P. sublimis*	Öpik (1930)	1	10.0	–
		P. cf. *P. sublimis*	Williams (1963)	–	–	–
		P. aff. *P. sublimis*	This study	5	2.7	1.6–3.6
D_valve	Ventral	*P. sublimis*	Öpik (1930)	1	4.2	–
		P. cf. *P. sublimis*	Williams (1963)	–	–	–
		P. aff. *P. sublimis*	This study	3	1.3	1.2–1.5
	Dorsal	*P. sublimis*	Öpik (1930)	1	5.8	–
		P. cf. *P. sublimis*	Williams (1963)	–	–	–
		P. aff. *P. sublimis*	This study	2	3.0	1.8–4.1

L, length; *W*, width; *D*, depth; hi, hinge. Valve depths of *P. sublimis* were calculated from Öpik (1930, pl. 5, fig. 50, left).

Table 17. Ratio means of *P. sublimis* (Öpik 1930), *P.* cf. *P. sublimis* (Williams 1963), and *P.* aff. *P. sublimis* (this study).

Ratio	Valve	Species	Data source	n	Mean (%)	Range (%)
L_valve × 100/W_valve	Ventral + dorsal	*P. sublimis*	Öpik (1930)	1	58	–
		P. cf. *P. sublimis*	Williams (1963)	–	67	–
		P. aff. *P. sublimis*	This study	15	70	52–91
W_hi × 100/W_valve	Ventral + dorsal	*P. sublimis*	Öpik (1930)	1	83	–
		P. cf. *P. sublimis*	Williams (1963)	–	–	–
		P. aff. *P. sublimis*	This study	5	94	89–95
D_valve × 100/L_valve	Ventral	*P. sublimis*	Öpik (1930)	1	61	–
		P. cf. *P. sublimis*	Williams (1963)	–	50	–
		P. aff. *P. sublimis*	This study	2	33	31–35
	Dorsal	*P. sublimis*	Öpik (1930)	1	83	–
		P. cf. *P. sublimis*	Williams (1963)	–	–	–
		P. aff. *P. sublimis*	This study	2	39	39–40
D_valve × 100/W_valve	Ventral	*P. sublimis*	Öpik (1930)	1	35	–
		P. cf. *P. sublimis*	Williams (1963)	–	–	–
		P. aff. *P. sublimis*	This study	2	23	20–27
	Dorsal	*P. sublimis*	Öpik (1930)	1	48	–
		P. cf. *P. sublimis*	Williams (1963)	–	–	–
		P. aff. *P. sublimis*	This study	2	28	27–30
W_msc × 100/L_msc	Ventral	*P. sublimis*	Öpik (1930)	–	50	–
		P. cf. *P. sublimis*	Williams (1963)	–	–	–
		P. aff. *P. sublimis*	This study	3	49	39–64
L_msc × 100/L_valve	Ventral	*P. sublimis*	Öpik (1930)	–	33	–
		P. cf. *P. sublimis*	Williams (1963)	–	–	–
		P. aff. *P. sublimis*	This study	3	35	27–42

L, length; *W*, width; *D*, depth; hi, hinge; msc, muscle scar. Öpik (1930, p. 108) listed the dimensions of a pair of conjoined valves of *P. sublimis* from which the ratios, excluding the valve depths, were calculated. Valve depths of *P. sublimis* were calculated from Öpik (1930, pl. 5, fig. 50, left).

from the Dunabrattin form based on its more transverse and significantly more convex valves (Table 17). Öpik (1930) did not describe or figure any interiors.

The Dunabrattin and Estonian forms are coeval making it difficult to determine migrational patterns. They appear to be mutually more different morphologically than the Irish and the slightly younger Welsh form but more material of *P. sublimis* including interiors is needed to clarify this. The Welsh form probably evolved from the Irish form, which migrated from a basin/slope setting of the Leinster Terrane to a shallow-water setting above wave base (Fortey *et al.* 2000) in the North Welsh Basin, Avalonia, during the early Sandbian. The ecological change from a deep to shallow-water setting may account for the development of larger valves, and possibly changes in other characters as well, in the Welsh form.

Occurrence. – Dunabrattin Limestone Formation, Units I–III, Dunabrattin–Bunmahon area.

Suborder Dalmanellidina Moore, 1952

Superfamily Dalmanelloidea Schuchert, 1913

Family Dalmanellidae Schuchert, 1913

Subfamily Dalmanellinae Schuchert, 1913

Genus *Howellites* Bancroft, 1945

Type species (by original designation). – *Reserella* (*Howellites*) *striata* Bancroft, 1945, from the Allt Ddu Group, upper Sandbian (lowermost *TS.5b*– middle *TS.5c*), Craig y Gath, Bala, Gwynedd, Wales.

Howellites hibernicus n. sp.

Plate 8, figures 1–11, 13–15; Plate 10, figures 10–12

Derivation of name. – Alluding to the Irish origin of the species.

Holotype. – NHMUK PI BB35247, interior mould of dorsal valve, Tramore Limestone Formation (Pl. 8, fig. 1).

Material. – Total of 72 ventral valves and 68 dorsal valves.

Diagnosis. – Variably shaped *Howellites* species commonly subcircular and ventribiconvex with a finely branched radial ornamentation dominated by internally inserted costellae; ventral muscle scar subtriangular to subquadrate extending forwards for over 35% of the ventral valve length beyond the well-developed dental plates; dorsal interior with undifferentiated to bilobed cardinal process; brachiophore bases commonly straight to slightly divergent.

Description. – Transversely wider than long to longer than wide but commonly subquadrate or subcircular *Howellites* species; strongly ventribiconvex in young growth stages and weakly ventribiconvex to almost planar in mature stages; cardinal angles dominantly orthogonal; 4–9 mm long ($n = 140$) and 4–11 mm wide ($n = 140$) with maximum width commonly anterior to hingeline; ventral valve 89% as long as wide (61–120%, $n = 72$) and 20% as deep as long (6–37%, $n = 70$); dorsal valve evenly convex, 81% as long as wide (56–105%, $n = 68$) and 11% as deep as long (6–22%, $n = 66$) with shallow sulcus extending from the umbo and greatly widening towards the anterior margin; ventral interarea straight, apsacline and narrow, dorsal interarea

straight and anacline; ornamentation finely ramicostellate (Pl. 10, fig. 12) with counts per mm at 5 mm anterior to the umbo of 4 and 5 ribs for 7 and 5 dorsal valves, respectively, and 4, 5 and 6 ribs for 3, 8 and 2 ventral valves, respectively; secondary and tertiary costellae are commonly developed in all sectors of the dorsal valve with occasional quaternary costellae present in sectors III and IV, represented most frequently by $3a^-1^-a^-$ and occasionally by $3a^-1^-a°$, $4a^-1^-a^-$ and $4a^-1°1^-$; the rib pattern most commonly associated with the first costa at 5 mm or more anterior to the umbo is $1a^-1^-$, $1a^-$, $1b^-$ with $1a°$ occurring in only one valve of 13; a similar pattern occurs in sector II, with $2a°$ occurring in 2 of 13 valves; the pattern for sector III and IV are similar but the addition of quaternary ribs and more frequent externals affect the basic pattern with $3a°$ and $4a°$ occurring in five of 13 and eight of 11 valves, respectively.

Ventral interior (Pl. 10, fig. 10) with subtriangular to subquadrate muscle scar, highly variable in outline with diductor scars extending anteriorly of the adductors; muscle scar 126% as long as wide (61–193%, $n = 67$) and 35% as long as the valve length (24–49%, $n = 67$); teeth large with well-developed, slightly divergent, straight dental plates; vascula media divergent and branching but only weakly impressed or obscured.

Dorsal interior (Pl. 10, fig. 11) with well-defined cardinalia; cardinal process varying from undifferentiated to bilobed, stout oval to narrow with parallel sides; narrow ancillary struts occasionally present; brachiophores short with bases commonly thickened by secondary shell accretion giving the them a subtriangular cross-section, the brachiophore bases vary from slightly convergent to slightly divergent with 4, 17 and 12 specimens being convergent, straight and divergent, respectively; discrete fulcral plates present in eight of 52 valves; dorsal adductor scar elongate, quadripartite with a faint division between the anterior and posterior adductors, adductor scar 89% as long as wide (65–139%, $n = 63$) and 44% as long as the valve length (30–61%, $n = 63$); proximal part of vascula media and vascula myaria commonly impressed.

Remarks on the material. – The main problem is to determine if more than one species of the genus is present in view of the great range of variation in all the calculated ratios. The material studied here mainly derive from sample A from the lower part of the Tramore Limestone Formation (Unit 2), containing 12 ventral and 13 dorsal valves, and sample B from the middle part of the formation (Unit 3), containing 58 ventral and 54 dorsal valves. Sample A was collected from a calcareous sandstone deposited on a shallow shelf (DR 2 or shallow DR 3; Fig. 5) and sample B was collected from calcareous shale and siltstone deposited on the mid to outer shelf (deep DR 3 or DR 4). The specimens of sample A are on average larger than those of sample B, the ones of sample A being 5–9 mm long and 6–11 mm wide with a mean of 7 mm ($n = 28$) and 8 mm ($n = 28$), respectively, compared to the ones of sample B being 4–8 mm long and 4–9 mm wide with a mean of 5 mm ($n = 112$) and 6 mm ($n = 112$), respectively. Due to the great range of the ratios, the small amount of measurements of sample A makes it unfit for statistical comparison as it is not representative. The samples were instead compared visually in RMA plots together with the *Howellites* specimens not belonging to sample A and B to assess if the data points plot separately and, hence, belong to different species. The following variables were plotted against valve lengths: valve width (Fig. 37), dorsal valve depth (Fig. 38), ventral valve depth (Fig. 39), dorsal muscle scar length (Fig. 40), ventral muscle scar length (Fig. 41), dorsal muscle scar length:width ratio (Fig. 42) and ventral muscle scar length:width ratio (Fig. 43).

The data points of sample A and B plot together in all the RMAs indicating they belong to the same species. Sample B represents a population including most growth stages, whereas Sample A, which includes mostly larger valves, must have been sorted by current activity removing the smaller valves from the population. This is in accordance with deposition of sample A in a shallow-water, moderate-energy environment above fair weather wave base (DR 2 to shallow DR 3), and sample B in a deeper, more quiet environment between fair weather and storm wave base (deep DR 3 or DR 4).

Morphological variation. – The total assemblage of *H. hibernicus* n. sp. was investigated for allometric growth. Valve length and width (*L*_valve and *W*_valve, respectively), ventral and dorsal valve depth (*D*_valve) and ventral and dorsal muscle scar length and width (*L*_msc and *W*_msc, respectively) included a sufficient amount of measurements for statistical analysis.

Valve dimensions
Total assemblage – valve width relative to valve length. – The log-transformed RMA plot (Fig. 44) of the valve widths plotted against valve length indicates no breakpoints in the distribution of the data although the two longest valves are separated from the rest of the assemblage by a gap in valve lengths

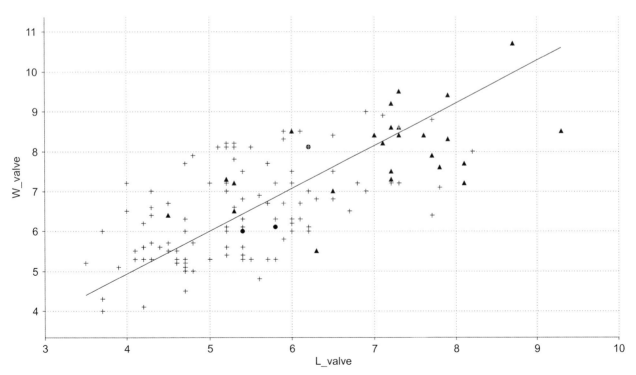

Fig. 37. Howellites hibernicus n. sp. RMA of the valve width plotted against the valve length for sample A (triangles), sample B (crosses) and other specimens (circles). $n = 140$, slope $a = 1.07$, error $a = 0.07$, $r = 0.68$, $P(\text{uncorr}) = 2 \times 10^{-20}$, significance level = 0.05. The data points plot together and there is no indication of the existence of more than one species. It is evident that sample A mostly includes adult specimens and, hence, was probably sorted by currents.

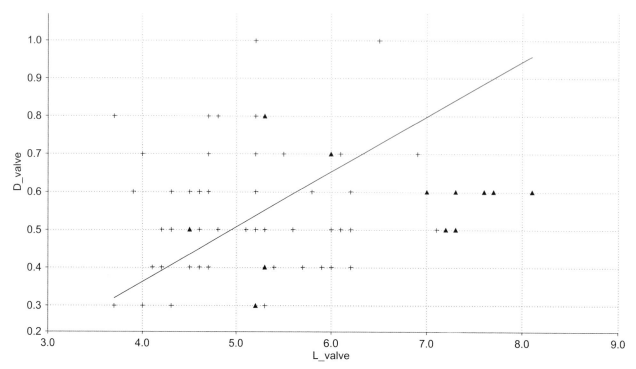

Fig. 38. Howellites hibernicus n. sp. RMA of the dorsal valve depth plotted against the valve length for samples A (triangles) and B (crosses). $n = 66$, slope $a = 0.15$, error $a = 0.02$, $r = 0.11$, P(uncorr) = 0.39, significance level = 0.05. The data points plot together and there is no indication of the existence of more than one species. The extremely scattered data points indicate that the dorsal valve depth was close to random compared to valve length.

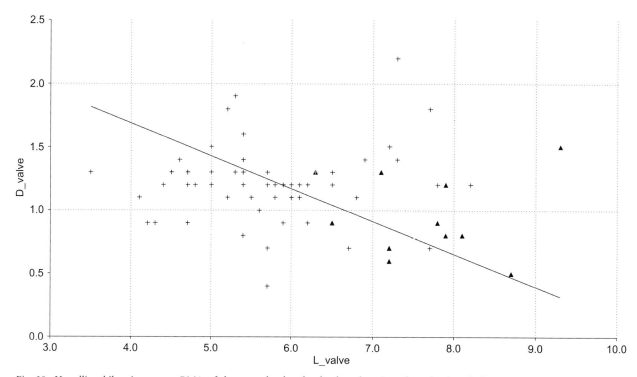

Fig. 39. Howellites hibernicus n. sp. RMA of the ventral valve depth plotted against the valve length for samples A (triangles) and B (crosses). $n = 70$, slope $a = -0.26$, error $a = 0.03$, $r = -0.14$, P(uncorr) = 0.24, significance level = 0.05. The data points plot together and there is no indication of the existence of more than one species. The ventral valves of sample A, however, plot within the less-convex end of the range of sample B.

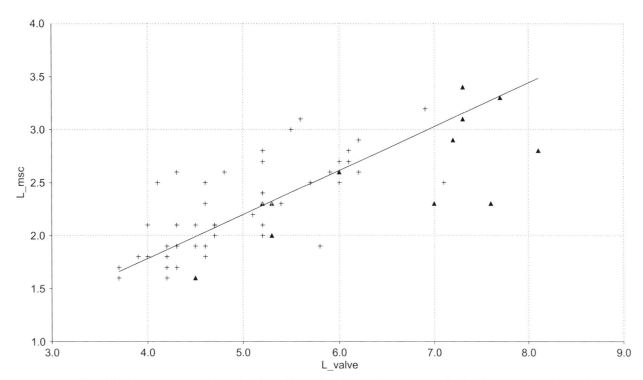

Fig. 40. *Howellites hibernicus* n. sp. RMA of the dorsal muscle scar length plotted against the valve length for samples A (triangles) and B (crosses). $n = 63$, slope $a = 0.42$, error $a = 0.03$, $r = 0.76$, $P(\text{uncorr}) = 8 \times 10^{-13}$, significance level = 0.05. The data points predominantly plot together, although the largest valves (all of them belonging to sample A) plot by themselves mainly beneath the regression line on the right side of the figure.

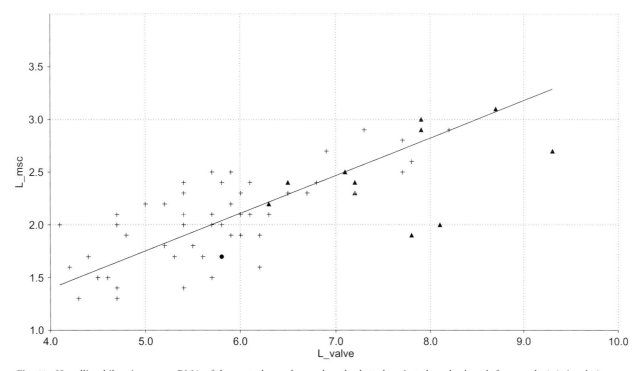

Fig. 41. *Howellites hibernicus* n. sp. RMA of the ventral muscle scar length plotted against the valve length for sample A (triangles), sample B (crosses) and one other specimen (circle). $n = 67$, slope $a = 0.36$, error $a = 0.03$, $r = 0.73$, $P(\text{uncorr}) = 2 \times 10^{-12}$, significance level = 0.05. The data points plot together and there is no indication of the existence of more than one species.

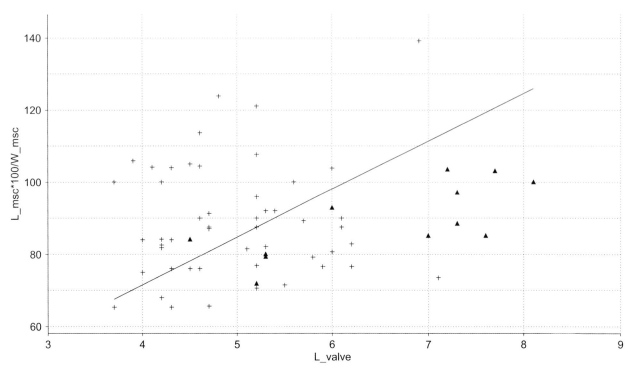

Fig. 42. Howellites hibernicus n. sp. RMA of the dorsal muscle scar length:width ratio ($L\_msc \times 100/W\_msc$) plotted against the valve length for samples A (triangles) and B (crosses). $n = 63$, slope $a = 13.26$, error $a = 1.68$, $r = 0.16$, P(uncorr) = 0.21, significance level = 0.05. The data points predominantly plot together, although the largest valves (all of them belonging to sample A) plot by themselves beneath the regression line on the right side of the figure.

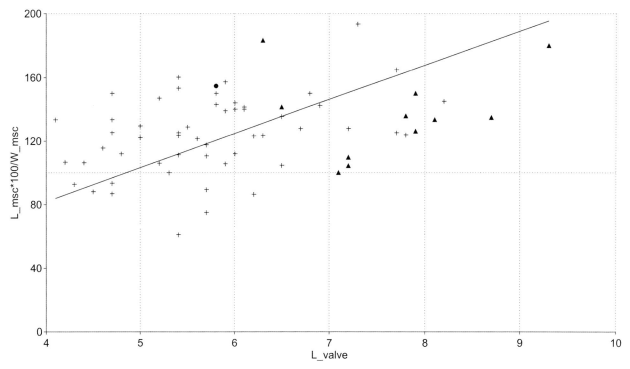

Fig. 43. Howellites hibernicus n. sp. RMA of the ventral muscle scar length:width ratio ($L\_msc \times 100/W\_msc$) plotted against the valve length for sample A (triangles), sample B (crosses) and one other specimen (circle). $n = 66$, slope $a = 21.24$, error $a = 2.44$, $r = 0.40$, P (uncorr) = 9×10^{-4}, significance level = 0.05. The data points plot together and there is no indication of the existence of more than one species.

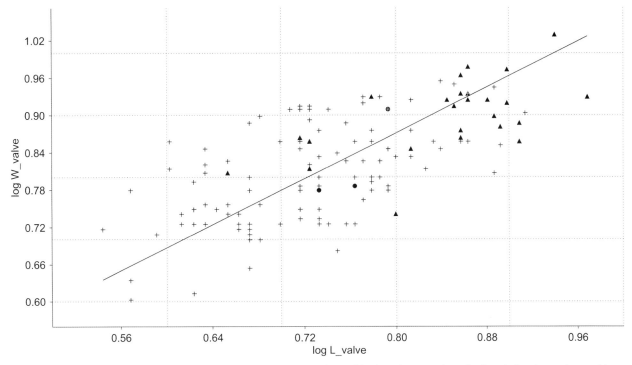

Fig. 44. Howellites hibernicus n. sp. Log-transformed RMA of the valve width plotted against the valve length for the total assemblage; triangles, sample A; crosses, sample B; circles, other specimens. $N = 140$, slope $a = 0.92$, error $a = 0.06$, $r = 0.68$, P(uncorr) $= 4 \times 10^{-20}$, $P(a = 1) = 0.19$, significance level $= 0.05$. Isometry cannot be rejected as $P(a = 1) > 0.05$ and the slope (a) is close to 1.

between 8.2 and 8.7 mm (~0.91 and 0.94 in the log-transformed RMA). Two specimens are not enough to divide the data into size-groups and analyse them against each other and they plot close to the regression line like the rest of the assemblage. The assemblage was therefore analysed statistically as one sample. The log-transformed RMA shows that length and width growth was isometric as P ($a = 1$) > 0.05 and the slope (a) is close to 1 and, hence, no changes in growth rate between these variables took place through ontogeny.

Dorsal valve depth relative to valve length. – The log-transformed RMA plot (Fig. 45) of the dorsal valve depths plotted against valve length indicates a possible breakpoint between valve lengths of 6.5 and 6.9 mm (~0.82 and 0.84 in the log-transformed RMA) where the larger valves plot beneath the regression line. The amount of data at hand for this group including the largest valves is too scarce for statistical comparison between the smaller and larger valves, however, due to the significant variation in the valve depths for a given valve length. Hence, it was not possible to assess the data statistically for significant changes in growth rate as indicated by the breakpoint. The assemblage was therefore analysed statistically as one sample for general allometric trends. The log-transformed RMA shows that relative growth of

dorsal valve depth and length was allometric as P ($a = 1$) < 0.05. This is very uncertain, however, as the data points are extremely scattered making the slope of the regression line less significant. The result is supported, though, by the non-transformed RMA plot of the dorsal valve depth:length ratio analysed against valve length (Fig. 46). Allometry is indicated as the graph shows a development from moderately convex, small dorsal valves to almost planar, large valves. The trace of the data points furthermore changes direction between 6.5 and 6.9 mm and flattens out above the regression line at depth:length ratios of about 8% supporting the existence of a breakpoint and, hence, indicating that a change in growth rates took place at this growth stage.

Ventral valve depth relative to valve length. – The log-transformed RMA plot (Fig. 47) of the ventral valve depths plotted against valve length indicates no breakpoints, although the two longest valves are separated from the rest of the assemblage by a gap in valve lengths between 8.2 and 8.7 mm as in the analysis of the valve lengths and widths. Like the former analysis, the assemblage had to be treated statistically as one sample. The log-transformed RMA shows that relative growth of ventral valve depth and length was allometric as $P(a = 1)$ < 0.05; however, this result is uncertain due to the very scattered data

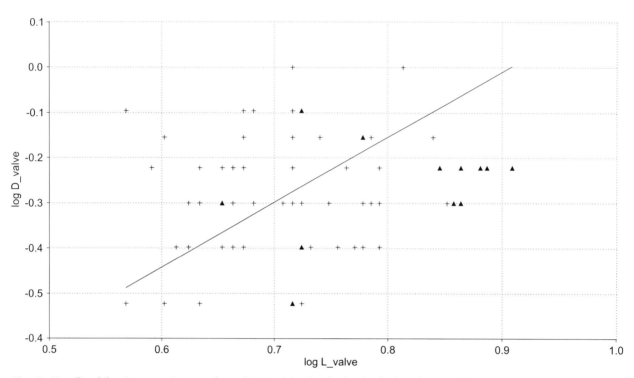

Fig. 45. Howellites hibernicus n. sp. Log-transformed RMA of the dorsal valve depth plotted against the valve length for sample A (triangles) and sample B (crosses). $n = 66$, slope $a = 1.44$, error $a = 0.18$, $r = 0.15$, P(uncorr) $= 0.22$, $P(a = 1) = 0.02$, significance level $= 0.05$. Allometric growth is indicated as $P(a = 1) < 0.05$. This is uncertain, however, as the data points are extremely scattered.

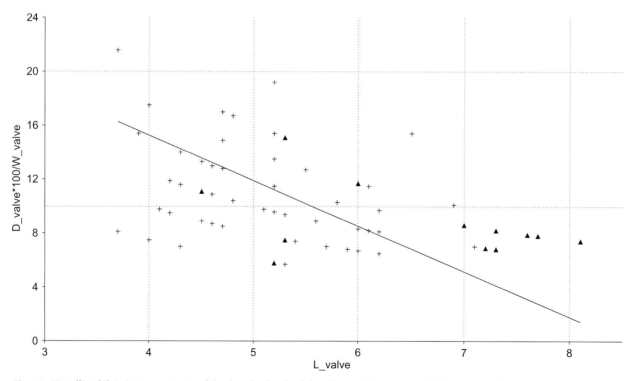

Fig. 46. Howellites hibernicus n. sp. RMA of the dorsal valve depth:length ratio (D_valve $\times$ $100/L$_valve) plotted against the valve length for sample A (triangles) and sample B (crosses). $n = 66$, slope $a = -3.38$, error $a = 0.37$, $r = -0.47$, P(uncorr) $= 8 \times 10^{-5}$, significance level $= 0.05$. Allometry is indicated as the graph show a change from moderately convex, small dorsal valves to almost planar, large valves through ontogeny and the trace of the data points flattens out, leaving the regression line to become constant around a valve depth of 8% the valve length for the largest valves.

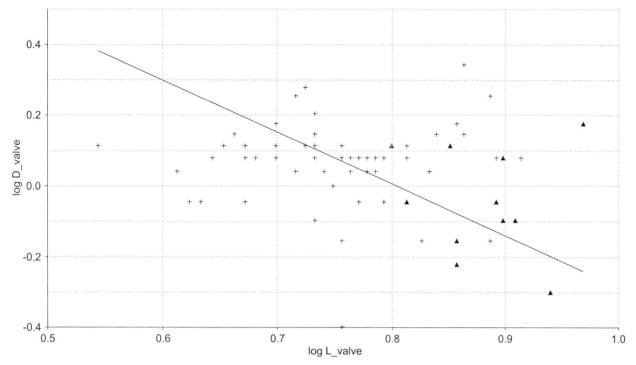

Fig. 47. Howellites hibernicus n. sp. Log-transformed RMA of the ventral valve depth plotted against the valve length for sample A (triangles) and sample B (crosses). $n = 70$, slope $a = -1.47$, error $a = 0.17$, $r = -0.20$, P(uncorr) = 0.10, $P(a = 1) = 6 \times 10^{-22}$, significance level = 0.05. Allometric growth is indicated as $P(a = 1) < 0.05$.

points making the slope of the regression line less significant. Indications of allometry are supported by the non-transformed RMA plot of the ventral valve depth:length ratio analysed against valve length (Fig. 48). It indicates a development from strongly convex ventral valves in immature specimens to almost planar valves in the largest, mature specimens. Despite the development of less convex ventral valves during ontogeny, their convexity remained greater than the convexity of the dorsal valves with corresponding valve lengths.

Dorsal muscle scar dimensions
Muscle scar length:width ratio relative to valve length. – The log-transformed RMA plot (Fig. 49) of the dorsal muscle scar length:width ratios plotted against valve length indicates a possible breakpoint between valve lengths of 6.2 and 6.9 mm (~0.82 and ~0.84 in the log-transformed RMA), where the larger valves primarily plot beneath the regression line. The number of valves equal to or longer than 6.9 mm is too few for statistical comparison of changing growth rates between the smaller and larger growth stages due to the significant variation in the dorsal muscle scar length:width ratios. Hence, the assemblage was analysed statistically as one sample for general allometric trends. The result of the log-transformed RMA is inconclusive as $P(a = 1) = 0.05$ and the data points are extremely scattered. The relation

between dorsal muscle scar length:width ratio and valve length in the non-transformed RMA plot in Figure 42 appears to be more or less random although there is a weak tendency of smaller valves having slightly wider muscle scars relative to muscle scar length (lower ratios).

Muscle scar length relative to valve length. – The log-transformed RMA plot (Fig. 50) of the dorsal muscle scar lengths plotted against valve length indicates a possible breakpoint between valve lengths of 6.2 and 6.9 mm (~0.82 and ~0.84 in the log-transformed RMA) where the larger valves mainly plot beneath the regression line. The amount of data at hand for this group including the largest valves of the assemblage is too sparse for statistical comparison and the assemblage was analysed statistically as one sample for general allometric trends. The log-transformed RMA indicates that the ventral muscle scar length increased isometrically relative to valve length through ontogeny as $P(a = 1) > 0.05$ and the slope (a) is close to 1.

Ventral muscle scar dimensions
Muscle scar length:width ratio relative to valve length. – The log-transformed RMA plot (Fig. 51) of the ventral muscle scar length:width ratios plotted against valve length indicates no breakpoints and no gaps in valve lengths dividing the data points into size-

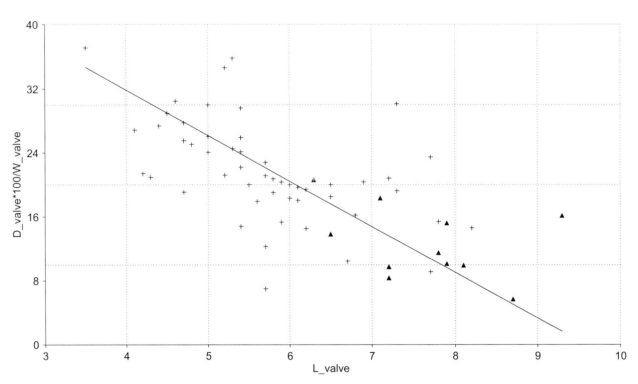

Fig. 48. *Howellites hibernicus* n. sp. RMA of the ventral valve depth:length ratio (D_valve $\times$ 100/L_valve) plotted against the valve length for sample A (triangles) and sample B (crosses). n = 70, slope a = −5.68, error a = 0.50, r = −0.68, P(uncorr) = 9 $\times$ 10^{-11}, significance level = 0.05. Allometry is indicated as the data points show a development from moderately convex small ventral valves to almost planar large valves through ontogeny.

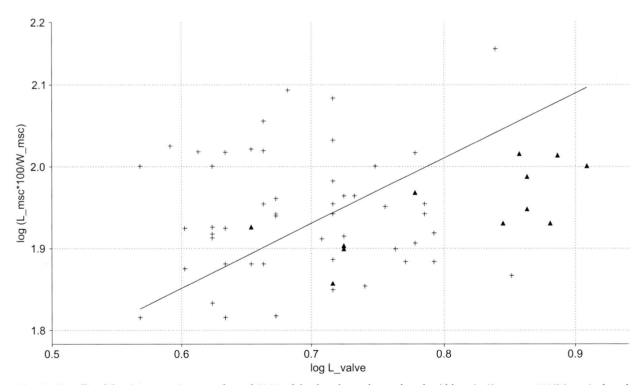

Fig. 49. *Howellites hibernicus* n. sp. Log-transformed RMA of the dorsal muscle scar length:width ratio (L_msc $\times$ 100/W_msc) plotted against the valve length for sample A (triangles) and sample B (crosses). n = 63, slope a = 0.80, error a = 0.10, r = 0.17, P(uncorr) = 0.19, P(a = 1) = 0.05, significance level = 0.05. The results are inconclusive as P(a = 1) = 0.05 and the data points are extremely scattered.

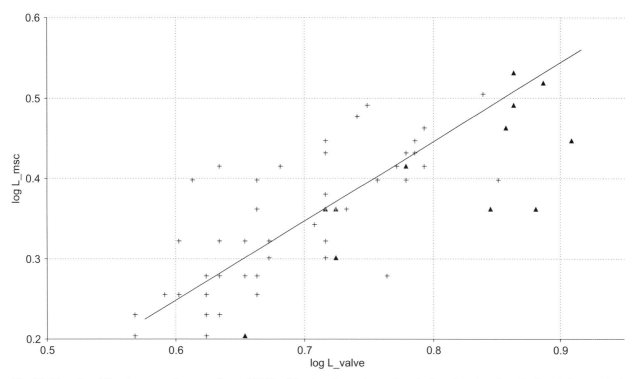

Fig. 50. Howellites hibernicus n. sp. Log-transformed RMA of the dorsal muscle scar length plotted against the valve length for sample A (triangles) and sample B (crosses). $n = 63$, slope $a = 0.99$, error $a = 0.08$, $r = 0.77$, $P(\text{uncorr}) = 2 \times 10^{-13}$, $P(a = 1) = 0.86$, significance level = 0.05. Isometry cannot be rejected as $P(a = 1) > 0.05$ and the slope (a) is close to 1.

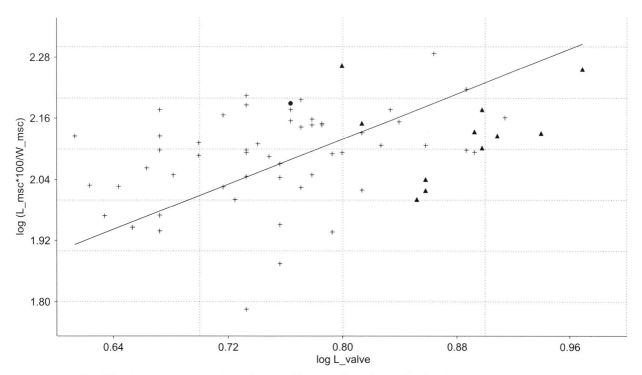

Fig. 51. Howellites hibernicus n. sp. Log-transformed RMA of the ventral muscle scar length:width ratio ($L\_msc \times 100/W\_msc$) plotted against the valve length for sample A (triangles), sample B (crosses) and one other specimen (circle). $n = 67$, slope $a = 1.10$, error $a = 0.13$, $r = 0.38$, $P(\text{uncorr}) = 2 \times 10^{-3}$, $P(a = 1) = 0.42$, significance level = 0.05. Isometry cannot be rejected as $P(a = 1) > 0.05$ and the slope (a) is close to 1.

groups and the assemblage was analysed statistically as one sample. The log-transformed RMA shows that the ventral muscle scar length:width ratio and the valve length increased isometrically through ontogeny as P ($a = 1$) > 0.05 and the slope (a) is close to 1. The relation between the ventral muscle scar length:width ratio and valve length appears to be more or less random like the dorsal muscle scar length:width ratio and dorsal valve length, with a comparable trend of smaller valves having slightly wider muscle scars relative to muscle scar length (lower ratios) (see non-transformed RMA plot; Fig. 43).

Muscle scar length relative to valve length. – The log-transformed RMA plot (Fig. 52) of the ventral muscle scar lengths plotted against valve length indicates no breakpoints and no gaps in valve lengths and the assemblage was analysed statistically as one sample. The log-transformed RMA indicates that the ventral muscle scar length increased isometrically relative to valve length through ontogeny as $P(a = 1)$ > 0.05 and the slope (a) is close to 1.

Summary
The analyses indicate that only the ventral and dorsal valve depth changed growth rate relative to valve length through ontogeny with both valves evolving from more convex in immature growth stages to

almost planar in mature growth stages. Despite this development, the ventral valves maintained greater convexities than the dorsal valves of corresponding valve lengths even for the largest growth stages.

Discussion. – The division of the family Dalmanellidae into genera cannot be based solely on presence or absence of certain features such as fulcral plates and crural fossettes or on convergence or divergence of the brachiophores. This was demonstrated by Williams & Wright (1963) who provided an estimate of the degree of variation in some of the genera, particularly *Howellites*, *Dalmanella* and *Onniella*. The variation existing in the specimens collected from the Tramore area suggests that they belong to the genus *Howellites*. The brachiophore bases vary from convergent to divergent but are predominantly straight to divergent and fulcral plates are rarely present; convexity and rib counts are also diagnostic of *Howellites*.

The previously recorded species assigned to *Howellites* from the British area were collected in Wales and are of late Sandbian age and younger (see Cocks 2008). Parkes (1994) noted that a mid–late Sandbian (*TS*.5b) dalmanellid, probably belonging to *Howellites*, dominated his new collections from Cloioge Upper and was present in samples from Raheen, southeast Ireland, but the specimens were indeterminate due to strong deformation. *Howellites*

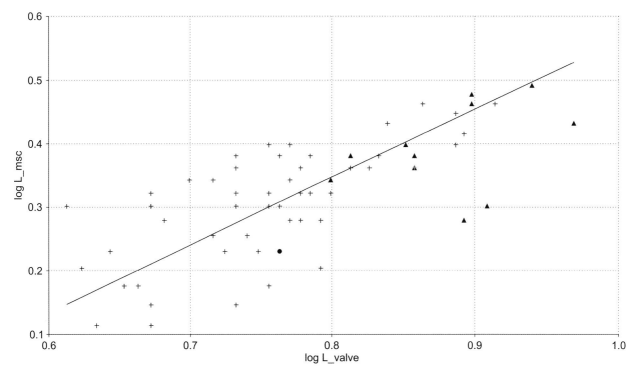

Fig. 52. *Howellites hibernicus* n. sp. Log-transformed RMA of the ventral muscle scar length plotted against the valve length for sample A (triangles), sample B (crosses) and one other specimen (circle). $n = 67$, slope $a = 1.07$, error $a = 0.09$, $r = 0.72$, P(uncorr) $= 1 \times 10^{-11}$, $P(a = 1) = 0.46$, significance level $= 0.05$. Isometry cannot be rejected as $P(a = 1)$ > 0.05 and the slope (a) is close to 1.

hibernicus n. sp. is the oldest recorded species of *Howellites* reported from the British and Irish area and it is below compared to the only other known Darriwilian species, *H. hammanni* Villas, 1985, from Spain as well as the morphologically more closely similar late Sandbian Welsh species of *Howellites*.

The ribbing associations (not the rib counts) of *Howellites* is the only character, which has proven to be consistently species diagnostic within this genus as shown by Williams (1963). He analysed several species and showed that relative growth in most cases was not species specific, which is supported by this study. The valve ratios commonly used to describe and identify brachiopod species are extremely variable within *Howellites*, in some cases due to allometric growth. Selected ratios of the Tramore form and the species described by Whittington (1938), Williams (1963) and Villas (1985) are provided in Table 18 for comparison. Although some of the mean values for a given ratio are very different between the species, the ranges often overlap making it difficult to identify a single (or a few) specimens using these variables alone. The ventral muscle scar length:width ratio for *H. hibernicus* n. sp. is probably diagnostic for this species as even the smallest ratio value is significantly larger than the largest values of the compared species (Table 18). The rib counts are not species diagnostic and all the compared species commonly have four or five primary costae with a mode of four per mm at 5 mm anterior to the umbo on the dorsal valves (Table 19). The analysis of *H. cruralis* (Whittington 1938) only included one specimen with five primary costae and, hence, it was not possible to give a modal value. Villas (1985) only provided the dominant number of ribs of *H. hammanni* and these are marked with an 'x' in the table. The diagnostic ribbing associations of *H. hibernicus* n. sp. are illustrated in Figure 53 and compared to the selected species in Table 20. None of the species compared with the Tramore form had ribbing associations similar to this as described in detail below, and, hence, it is considered a new species.

Howellites hammanni was described by Villas (1985) from the Alpartir and Sierra members of the Castillejo Formation, upper Darriwilian – lower Sandbian, Fombuena area, Iberian Chains, Spain. It is the only other known *Howellites* contemporary with *H. hibernicus* n. sp. in the Darriwilian but it is morphologically very different. *Howellites hammanni* is larger with a maximum length and width of 13.6 mm and 18.1 mm, respectively (see Villas 1985; p. 63). The length:width ratio of *H. hammanni* is considerably smaller than for the Tramore

form as well as for any of the other species compared to it (Table 18) and it may in this case be species diagnostic. However, given the great variation within the other species, more material of *H. hammanni* is needed to confirm this. *Howellites hammanni* has the same dominant number of primary costae (4–5) per mm at 5 mm anterior to the umbo as the Tramore form and the other discussed species (Table 19), but it appears to be very different in its dorsal ribbing associations, and the frequency of external costellae in sector I–III varies with the depth of the sulcus (Villas 1985). Villas (1985, pl. 12, fig. 9 and pl. 13, fig. 2) figured two dorsal valves, allowing investigating of the ribbing associations. Both ribs were present in the association including $1a^-1^-$ and $1b^-$; however, these were inserted in the opposite order than in *H. hibernicus* n. sp. with $1b^-$ inserted just before $1a^-1^-$ on both specimens. Either one or both ribs were absent in the other four compared associations (Table 20).

Williams (1963, p. 404, table 53) provided ribbing associations of the following species: *H. antiquior* (M'Coy, 1852 *in* Sedwick & M'Coy 1851-55); *H. ultimus* Bancroft, 1945; *H.* cf. *H. ultimus* Bancroft, 1945; *H. intermedius* Bancroft, 1945; *H. striatus* Bancroft, 1945; and *H.* cf. *H. striatus* Bancroft, 1945, from the upper Sandbian (earliest *TS.*5b–mid *TS.*5c) of North Wales. *Howellites* cf. *ultimus* and *H.* cf. *striatus* have rib associations comparable to *H. ultimus* and *H. striatus*, respectively, and, hence, are not included in Table 20. The ribbing association of *H. hibernicus* n. sp. is very distinct from all of these as it only includes three of the five associations observed in the Welsh species, together with different proportions of inserted costellae for most of the present rib associations (Table 20).

Howellites cruralis was described by Whittington (1938) from beds of late Sandbian age (lowermost *TS.*5b–middle *TS.*5c) at Powys, South Wales. Whittington did not quantify his observations and this study measured his syntypes (pl. 10, figs 8–11) for comparison (Table 18). Good quality pictures of his syntypes are available from the online catalogue of the Lapworth Museum of Geology: http://mimsy.bham.ac.uk/. It was not possible to investigate the rib associations of sector IV due to poor preservation; however, $1b^-$ and $3a°$ were inserted before $1a^-1^-$ and $3c^-$, respectively, which is in the opposite order than for *H. hibernicus* n. sp. (Table 19).

Occurrence. – Tramore Limestone Formation, Units 1–3 and 5, Barrel Strand area. Dunabrattin Limestone Formation, Units I–IV, Dunabrattin–Bunmahon area.

Table 18. Ratios of selected species of *Howellites*.

Ratio	Valve	Species	Data source	*n*	Mean (%)	Range (%)
$L\_valve \times 100/W\_valve$	Ventral	*H. antiquior*	Williams (1963)	4	111	81–163
		H. cruralis	Whittington (1938)	1	114	–
		H. hammanni	Villas (1985)	4	69	60–76
		H. hibernicus n. sp.	This study	69	89	61–120
		H. intermedius	Williams (1963)	2	94	82–105
		H. striatus	Williams (1963)	3	102	95–110
		H. ultimus	Williams (1963)	3	84	81–89
	Dorsal	*H. antiquior*	Williams (1963)	4	77	71–84
		H. cruralis	Whittington (1938)	3	94	83–100
		H. hammanni	Villas (1985)	5	61	53–71
		H. hibernicus n. sp.	This study	67	81	56–105
		H. intermedius	Williams (1963)	1	69	–
		H. striatus	Williams (1963)	5	80	75–88
		H. ultimus	Williams (1963)	3	79	76–81
$D\_valve \times 100/L\_valve$	Ventral	*H. antiquior*	Williams (1963)	–	>40	–
		H. cruralis	Whittington (1938)	–	–	–
		H. hammanni	Villas (1985)	–	–	–
		H. hibernicus n. sp.	This study	67	20	6–37
		H. intermedius	Williams (1963)	–	>33	–
		H. striatus	Williams (1963)	–	>33	–
		H. ultimus	Williams (1963)	–	>33	–
	Dorsal	*H. antiquior*	Williams (1963)	–	<14	–
		H. cruralis	Whittington (1938)	–	–	–
		H. hammanni	Villas (1985)	–	–	–
		H. hibernicus n. sp.	This study	65	11	6–22
		H. intermedius	Williams (1963)	–	13	–
		H. striatus	Williams (1963)	–	<14	–
		H. ultimus	Williams (1963)	–	17	–
$L\_msc \times 100/W\_msc$	Ventral	*H. antiquior*	Williams (1963)	–	<80	–
		H. cruralis	Whittington (1938)	1	98	–
		H. hammanni	Villas (1985)	5	89	67–130
		H. hibernicus n. sp.	This study	11	147	124–193
		H. intermedius	Williams (1963)	–	>80	–
		H. striatus	Williams (1963)	–	80	–
		H. ultimus	Williams (1963)	–	75	–
$L\_msc \times 100/L\_valve$	Ventral	*H. antiquior*	Williams (1963)	–	<40	–
		H. cruralis	Whittington (1938)	1	27	–
		H. hammanni	Villas (1985)	3	36	34–40
		H. hibernicus n. sp.	This study	11	33	24–40
		H. intermedius	Williams (1963)	–	>40	–
		H. striatus	Williams (1963)	–	40	–
		H. ultimus	Williams (1963)	–	40	–
	Dorsal	*H. antiquior*	Williams (1963)	–	>50	–
		H. cruralis	Whittington (1938)	2	33	31–34
		H. hammanni	Villas (1985)	–	>50	–
		H. hibernicus n. sp.	This study	3	43	43–44
		H. intermedius	Williams (1963)	62	44	30–61
		H. striatus	Williams (1963)	–	60	–
		H. ultimus	Williams (1963)	–	>60	–

L, length; *W*, width; *D*, depth; msc, muscle scar. Valve length:width ratios ($L\_valve \times 100/W\_valve$) of Williams' (1963) species were recalculated by measuring his figured specimens (*H. antiquior* pl. 6, figs 13–19, pl. 7, figs 1, 2,5, 6; *H. intermedius* pl. 6, figs 1–5; *H. striatus* pl. 5, figs 8–18; *H. ultimus* pl. 6, figs 6–12). Ratios of *Howellites cruralis* are calculated from Whittington (1938, figured specimens, pl. 10, figs 8–11). Ratios of *Howellites hammanni* are from Villas (1985, p. 63).

Family Paurorthidae Öpik, 1933b

Genus *Paurorthis* Schuchert & Cooper, 1931

Type species (by original designation). – *Orthambonites parva* Pander, 1830, from the Volkhov and Lynna formations, Dapingian–middle Darriwilian, northwest Russia.

Paurorthis aff. *P. parva* (Pander, 1830)

Plate 8, figures 16–23

aff. 1830   *Orthambonites parva* Pander, p. 83, pl. 26, figs 10a–c.

aff. 1932   *Paurorthis parva* (Pander); Schuchert & Cooper, p. 79, pl. 3, figs 5–8, 10.

Table 19. Number of costellae per mm at 5 mm anterior to the umbo on dorsal valves of selected *Howellites* species.

Species	Data source	3	4	5	6
H. antiquior	Williams (1963)	–	[24]	16	2
H. cruralis	Whittington (1938)	–	–	1	–
H. intermedius	Williams (1963)	–	[14]	7	2
H. hammanni	Villas (1985)	?	x	x	?
H. hibernicus n. sp.	This study	–	[7]	5	–
H. striatus	Williams (1963)	1	[27]	11	1
H. ultimus	Williams (1963)	3	[23]	18	–

Rib counts of *Howellites striatus* are from Williams (1963, data for *H. striatus*, *H. intermedia expectata*, and *H. striata lineata*). Rib counts of *Howellites antiquior* are from Williams (1963; for Meifod and Bala forms described by M'Coy 1852 *in* Sedwick & M'Coy 1851-55). 'x', common number of ribs/mm at 5 mm anterior to the umbo for *Howellites hammanni* (i.e. Villas 1985).

Table 20. Comparison of diagnostic ribbing associations between selected species of *Howellites*.

Species	Data source	1a⁻1⁻)1b⁻	2c⁻)2a°	3c⁻)3a°	4b⁻)4b°	4a⁻1°)4b⁻1⁻
H. antiquior	Williams (1963)	9)23	10)10	39)42 [2]	31)33 [4]	2)14 [1]
H. cruralis	Whittington (1938)	0)1	–	0)1	?	?
H. hammanni	Villas (1985)	0/2	–	–	–	–
H. hibernicus n. sp.	This study	6)0	0)3	0)8	9)1	1)2
H. intermedius	Williams (1963)	7)40	25)33	49)68	38)48 [4]	1)23 [1]
H. striatus	Williams (1963)	5)25 [5]	14)20	56)76 [4]	48)64 [11]	4)29 [4]
H. ultimus	Williams (1963)	11)24 [2]	13)20	19)67 [4]	29)64 [13]	17)23 [2]

In a given association, the proportions show the number of dorsal valves in which a given costella is inserted before the other counting from the umbo towards the anterior margin of the valve. The number of valves in which the association occurred is given before the ')' and the number of valves in which the ribs were present but were not inserted in the specified order is given after the ')'. The number of valves in which both costellae originated at the same growth stage is given in square brackets. Cases where one or both ribs were absent is marked by a '–'. The rib associations of *H. striatus* include the data given by Williams (1963) for *H. striatus*, *H. intermedia expectata*, and *H. striata lineata* as these are synonymous (see Cocks 2008; p. 149). The rib associations of *H. antiquior* include the data given by Williams (1963) for both the Meifod and Bala forms described by M'Coy (1852 *in* Sedwick & M'Coy 1851-55).

aff. 1933b *Paurorthis parva* (Pander); Öpik, p. 12, pl. 3, 4, 6, fig. 4.

aff. 1956   *Paurorthis parva* (Pander); Cooper, p. 971, pl. 151, D, fig. 22.

aff. 1961   *Paurorthis parva* (Pander); Rubel, p. 196, pl. 12, figs 1–10, text-figs 13–16.

aff. 2000   *Paurorthis parva* (Pander); Harper, p. 810, fig. 587, 1a–d.

aff. 2004   *Paurorthis parva* (Pander); Egerquist, pl. 4, figs 9a–d.

Material. – Total of seven ventral valves and eight dorsal valves.

Description. – Subcircular, ventribiconvex *Paurorthis* species with maximum width medially, obtuse cardinal angles and evenly rounded lateral and anterior margins; length 6–11 mm (*n* = 15) and width 7–12 mm (*n* = 15); ventral hingeline 72% as wide as the maximum width (63–88%, *n* = 4) and dorsal hingeline 66% as wide as the maximum width (*n* = 1); ventral valve evenly convex with maximum depth medially, about as long as wide (88–114%,

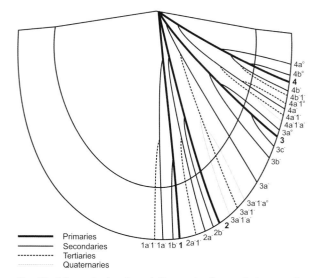

Fig. 53. Schematic drawing of diagnostic rib associations on the dorsal valves of *Howellites hibernicus* n. sp.

n = 7) and 37% as deep as long (*n* = 1); dorsal valve evenly convex, 80% as long as wide (66–90%, *n* = 8) and 25% as deep as long (23–28%, *n* = 2) with a gentle sulcus becoming wider and shallower anteriorly; ventral interarea strongly curved, apsacline, 16% as long as the valve length (15–19%, *n* = 6), delthyrium partly closed by variably developed lateral deltidial plates; dorsal interarea short, anacline, with open notothyrium; ornament fascicostellate with gently elevated, well-defined fascicles commonly composed of three ridged rather than rounded costellae, the number of costella may increase from three to five before the primary fascicles divide into two; costellae number four and five per mm at 5 mm anteromedially to the umbo on two ventral and one dorsal valve, respectively; on one dorsal exterior two primary fascicles occur in the sulcus at 4 mm anterior to the umbo and these divide between 4 and 6 mm to give four fascicles at 6–10 mm.

Ventral interior with short, stout, divergent teeth supported by dental plates greatly thickened by secondary shell; crural fossettes deeply impressed on the internal tooth surface; ventral muscle scar elongately subtriangular, rounded anteriorly, 56% as wide as long (49–63%, *n* = 6) and 36% as long as the valve length (33–40%, *n* = 6); adductors and diductors obscured due to poor preservation; a shallow median ridge anterior of the ventral muscle scar is developed in two of six specimens; mantle canal system poorly impressed.

Dorsal interior with fine ridgelike cardinal process on a raised notothyrial platform, brachiophores strong, bladelike and divergent, subtending an angle of about 90°, joined to notothyrial platform at their bases where they may be greatly thickened by secondary shell, 45% as long as their lateral spread (38–52%, *n* = 6) and 22% as long as the valve length (20–24%, *n* = 6); median ridge rounded and continuous with the notothyrial platform, bisecting the adductor scar which is poorly impressed and weakly quadripartite with narrow, subrectangular posterior adductors and large, suboval anterior adductors; adductor scar 70% as long as wide (61–78%, *n* = 3) and 41% as long as the valve length (36–45%, *n* = 3); mantle canal system weakly impressed or obscured.

Discussion. – The *Paurorthis* specimens from Tramore are similar to the Baltic species *P. parva* Pander, 1830, described from northwest Russia and north Estonia.

The *2008–2014 SARV: Geoscience collections and data repository* currently includes 216 records of Dapingian–Darriwilian *P. parva* from north Estonia and northwest Russia with 42 photos of 26 different valves and is available on the Geoscience Collections of Estonia home page http://geokogud.info/search. php. Two of the 216 occurrences are erroneously correlated with the Kukruse Stage (uppermost Darriwilian). Most of these figured specimens as well as the late Dapingian – early Darriwilian specimens from northwest Russia illustrated by Schuchert & Cooper (1932, pl. 3, figs 5, 7, 8, 10) and refigured by Harper (2000, fig. 587, 1a–d) for the *Treatise on Invertebrate Paleontology* were measured for comparison with the Tramore specimens. Selected ratios used for taxonomical description were calculated from the measurements, and the results are provided in Table 21 and rib counts in Table 22. The measured specimens are within the same size range and have similar rib counts but the mean values and the ranges of the calculated ratios differ slightly in most cases indicating some degree of evolutionary change between *P. parva* and the Tramore species. They are morphologically closely related however, also indicated by the similar expression of their qualitative data. All of the compared specimens are subcircular and ventribiconvex; the ventral valves have stout teeth with heavily thickened dental plates and prominent crural fossettes; the dorsal valves possess a shallow sulcus widening anteriorly, their brachiophores diverge at about 90° and join at their thickened bases with a well-elevated notothyrial platform. When comparing the specimens from Russia, Estonia and Tramore, it is apparent that the distinctness of the growth lines and the thickness of the ventral median ridge (when present) are variable.

Williams (1962) described *Paurorthis* sp. from the middle Sandbian (lower *TS*.5b) Balclatchie Formation and upper Sandbian – lower Katian (upper *TS*.5b–upper middle *TS*.5c) Ardwell Farm Formation, Girvan, Scotland, which appears to be rather similar to *P.* aff. *P. parva* from Tramore. The Scottish form is comparable in ventral valve convexity with ventral valves being 40% as deep as long, in relative dorsal valve dimensions being 80% as long as wide, and it has similar rib counts of 4–5 costellae per mm at 5 mm anteromedially to the umbo. They are also similar in their ventral muscle scar length:width ratio the muscle scar of *Paurorthis* sp. being 33% as long as wide and they have similar dorsal interiors. Williams noted a high wedge of secondary shell was inserted immediately anterior to the ventral muscle scar and noted it was unusual compared to other species of *Paurorthis* but this falls within the morphological range of the specimens investigated by this study. It was not possible to assess the details of the fasciculation from Williams' figured specimens (pl. 12, figs 41, 42) but it resembles *P.* aff. *P. parva* in all other details and may be conspecific, although Williams compared it to *Paurorthis* aff. *P. fasciculata* Cooper, 1956. He noted some important differences between his form

Table 21. Selected ratios of *Paurorthis* aff. *P. parva* from Tramore and *Paurorthis parva* from north Estonia and northwest Russia.

Ratio	Valve	Data source	n	Mean (%)	Range (%)
$L\_valve \times 100/W\_valve$	Ventral	This study	7	100	88–114
		Harper (2000)	2	91	–
		2008–2014 SARV	7	95	88–100
	Dorsal	This study	8	80	66–90
		Harper (2000)	2	82	79–85
		2008–2014 SARV	12	87	77–95
$W\_hi \times 100/W\_valve$	Ventral	This study	4	72	63–88
		Harper (2000)	2	67	62–73
		2008–2014 SARV	3	74	73–76
	Dorsal	This study	1	66	–
		Harper (2000)	2	67	67–68
		2008–2014 SARV	12	73	58–84
$L\_int \times 100/L\_valve$	Ventral	This study	4	16	15–19
		Harper (2000)	1	15	–
		2008–2014 SARV	–	–	–
$W\_msc \times 100/L\_msc$	Ventral	This study	6	56	49–63
		Harper (2000)	1	56	–
		2008–2014 SARV	5	62	52–72
$L\_msc \times 100/W\_msc$	Dorsal	This study	3	70	61–78
		Harper (2000)	1	67	–
		2008–2014 SARV	7	73	56–81
$L\_msc \times 100/L\_valve$	Ventral	This study	6	36	33–40
		Harper (2000)	1	37	–
		2008–2014 SARV	5	43	37–46
	Dorsal	This study	3	41	36–45
		Harper (2000)	1	39	–
		2008–2014 SARV	7	37	26–44
$L\_br \times 100/W\_br$	Dorsal	This study	6	45	38–52
		Harper (2000)	1	37	–
		2008–2014 SARV	8	48	42–56
L_br/L_valve	Dorsal	This study	6	22	20–24
		Harper (2000)	1	20	–
		2008–2014 SARV	8	21	18–25

L, length; *W*, width; hi, hingeline; int, interarea; msc, muscle scar; br, brachiophores. The ratios of the Baltic *P. parva* specimens were calculated from illustrated specimens (Harper 2000; *2008–2014 SARV: Geoscience collections and data repository*).

Table 22. Rib counts of *Paurorthis* aff. *P. parva* from Tramore and *Paurorthis parva* from north Estonia and northwest Russia.

Character	Valve	Data source	n	Mean	Range
No. of costellae in primary fascicles before division	Ventral + dorsal	This study	3	3	3–5
		Harper (2000)	2	3	3–5
		2008–2014 SARV	13	3	3–5
No. of costellae per mm at 5 mm growth stage	Ventral + dorsal	This study	2	4	–
		Harper (2000)	1	4	4–5
		2008–2014 SARV	5	4	4–5
		This study	1	5	–
		Harper (2000)	1	5	–
		2008–2014 SARV	9	5	4–5
No. of primary fascicles in sulcus at 4 mm growth stage	Dorsal	This study	1	2	–
		Harper (2000)	1	2	–
		2008–2014 SARV	9	2	–
No. of fascicles in sulcus at 6–10 mm growth stage	Dorsal	This study	1	4	–
		Harper (2000)	1	4	–
		2008–2014 SARV	9	4	–

Counts of ribs of Baltic *P. parva* specimens are from illustrated specimens (Harper 2000; *2008–2014 SARV: Geoscience collections and data repository*).

and *P.* aff. *P. fasciculata*, though, such as a coarser ornamentation and the presence of the wedge extending anteriorly from the ventral muscle scar in his form.

Occurrence. – Tramore Limestone Formation, Units 1–3, Barrel Strand area and Quillia Section 2. Dunabrattin Limestone Formation, Units I–III, Dunabrattin–Bunmahon area.

Superfamily Enteletoidea Waagen, 1884
Family Linoporellidae Schuchert & Cooper, 1931

Genus *Salopia* Williams *in* Whittington & Williams, 1955

Type species (by original designation). – *Orthis salteri* Davidson, 1869, from the Horderley Sandstone Formation, upper Sandbian (lowermost *TS.*5b–middle *TS.*5c), Horderley, Shropshire, England.

Salopia gracilis Williams *in* Whittington & Williams, 1955

Plate 8, figure 12; Plate 9, figures 1–9

1955   *Salopia salteri gracilis* Williams *in* Whittington & Williams, p. 410, pl. 38, figs 47–51.

1978   *Salopia salteri gracilis* Williams; Cocks, p. 82.

2008   *Salopia salteri gracilis* Williams; Cocks, p. 171.

Material. – Total of nine ventral valves and 17 dorsal valves.

Description. – Subcircular, ventribiconvex *Salopia* species with obtuse cardinal angles slightly greater than 90°; lateral and anterior margins curved; maximum width just anterior to hingeline; length 4–12 mm (*n* = 26) and width 6–14 mm (*n* = 25); ventral valve evenly convex with maximum depth posteromedially, 91% as long as wide (69–122%, *n* = 8) and 33% as deep as long (*n* = 1); dorsal valve evenly convex and weakly sulcate, 79% as long as wide (56–95%, *n* = 17) and 18% as deep as long (12–25%, *n* = 2); ventral interarea long, curved, apsacline with open delthyrium, 25% as long as the valve length (23–29%, *n* = 3); dorsal interarea shorter, anacline with open notothyrium; ornament finely multicostellate with faint concentric growth lines, 7–8 costellae per mm at 5 mm anterior to the umbo counted on two ventral valves, and a total of about 70 and 80 costellae counted on one dorsal and one ventral valve, respectively.

Ventral interior with simple teeth supported by long dental plates which define the lateral margins of the muscle scar; ventral muscle scar suboval to elongately subtriangular with adductor scars not extending as far anteriorly as the lobate diductor scars, 64% as wide as long (53–89%, *n* = 8), 22% as wide as the valve width (17–29%, *n* = 8) and 39% as long as the valve length (34–41%, *n* = 8).

Dorsal interior with cardinalia 22% as long as the valve length (17–32%, *n* = 10) consisting of a simple ridgelike cardinal process extending forwards into a broad, high median septum, simple, bladelike brachiophores with tops subtending an angle of less than 90° and bases converging onto the median septum to form a septalium, and a pair of concave fulcral plates defining deep sockets laterally; median septum extending anteriorly for 69% of the valve length measured from the umbo (55–84%, *n* = 14); dorsal adductor scar quadripartite, bisected medially by the median septum and laterally by the vascula myaria, 86% as long as wide (65–108%, *n* = 10), 37% as wide as valve width (24–44%, *n* = 10) and 40% as long as the valve length (36–46%, *n* = 10); anterior adductor scars larger than the posterior scars; mantle canal system poorly impressed but some canals are observed peripherally.

Discussion. – The *Salopia* specimens from Tramore are regarded conspecific with *Salopia gracilis* Williams *in* Whittington & Williams, 1955; from the Derfel Limestone, middle Sandbian (middle *TS.*5b), Gwynedd, Wales. *Salopia* was first recorded from southeast Ireland by Parkes (1994) who described *Salopia* sp. from upper Sandbian rocks (lower *TS.*5c) of Kildare, Grange Hill, Horizon 2 and Greenville Moyne but the presence of *Salopia* in the lower part of the Tramore Limestone moves the first occurrence of this genus in southeast Ireland back to the latest Darriwilian.

Selected ratios used for taxonomical description were compared between the Tramore form, *S. gracilis* and *S. salteri* (Davidson, 1869) (Table 23). Williams *in* Whittington & Williams (1955) provided a list of mean values for a number of ratios of *S. gracilis* and *S. salteri*, but he did not include the ratio ranges nor measurements of the individual specimens. The size of Williams' figured *S. gracilis* specimens falls within the range of the Tramore specimens, their mean ratio values are almost identical (Table 23) and they have similar rib counts with a total about 70 costella. They also have very similar interior and exterior morphologies when comparing Williams' description and figured specimens with the Tramore form. The Tramore form appears to have slightly more well-defined adductor scars and a generally thicker and more rounded dorsal median septum. This may be due to ecological differences as the two assemblages originated in different environments. The

Table 23. Selected ratios of *Salopia gracilis* and *Salopia salteri*.

Ratio	Valve	Species	Data source	*n*	Mean %	Range %
$L\_valve \times 100/W\_valve$	Ventral	*S. gracilis* (SE Ireland)	This study	8	91	69–122
		S. gracilis (Wales)	Williams *in* Whittington & Williams (1955)	3	95	–
		S. salteri (England)	Williams *in* Whittington & Williams (1955)	–	–	–
$W\_msc \times 100/W\_valve$	Ventral	*S. gracilis* (SE Ireland)	This study	8	22	17–29
		S. gracilis (Wales)	Williams *in* Whittington & Williams (1955)	10	23	–
		S. salteri (England)	Williams *in* Whittington & Williams (1955)	4	34	–
$W\_msc \times 100/L\_msc$	Ventral	*S. gracilis* (SE Ireland)	This study	8	64	53–89
		S. gracilis (Wales)	Williams *in* Whittington & Williams (1955)	10	70	–
		S. salteri (England)	Williams *in* Whittington & Williams (1955)	4	100	–
$L\_int \times 100/L\_valve$	Ventral	*S. gracilis* (SE Ireland)	This study	3	25	23–29
		S. gracilis (Wales)	Williams *in* Whittington & Williams (1955)	10	27	–
		S. salteri (England)	Williams *in* Whittington & Williams (1955)	3	50	–
$L\_car \times 100/L\_valve$	Dorsal	*S. gracilis* (SE Ireland)	This study	10	22	17–32
		S. gracilis (Wales)	Williams *in* Whittington & Williams (1955)	10	20	–
		S. salteri (England)	Williams *in* Whittington & Williams (1955)	4	30	–

L, length; *W*, width; msc, muscle scar; int, interarea; car, cardinalia. Mean values for the Welsh *Salopia gracilis* and English *Salopia salteri* are from Williams *in* Whittington & Williams (1955).

Tramore samples were collected from calcareous sand- and siltstones deposited in shallow-water environments (DR 2–3; see Fig. 5) and the Welsh form was recorded from calcareous mudstones deposited on the more distal part of the shelf (probably DR 4). The differences may also be related to evolution as the Tramore form is slightly older than the Welsh form but the few differences are not considered enough to define a new species or subspecies.

Williams *in* Whittington & Williams (1955) suggested keeping *S. gracilis* as a subspecies of *S. salteri* as he believed that some of the differences he recorded between these two forms may have been controlled by ecology rather than being species diagnostic. He argued that growth of the interarea may depend on the nature of the seafloor, *S. salteri* being recorded from a pebbly sandstone and *S. gracilis* from calcareous mudstones. He also pointed out that the proportionate development of the muscle scars, dental plates and possibly the cardinalia may in turn be dependent on the size of the interarea. The ventral interarea length of *S. gracilis* from Tramore is identical to the Welsh *S. gracilis* specimens (Table 23) even though they lived on very different sediments. Moreover, no pronounced differences in ventral interarea length of the Tramore specimens are observed by this study even though they were collected from a range of fine- and coarse-grained calcareous siliciclastics. Based on this and the significantly wider ventral muscle scar relative to valve width and length as well as the significantly longer ventral interarea of *S. salteri* with means not even included in the ratio ranges of *S. gracilis* (Table 23), we prefer to recognize *S. gracilis* as a separate species.

Seven formally described *Salopia* species, including *S. gracilis* and *S. salteri*, have previously been recorded from the British Isles (Cocks 2008). The other five species are *S. abbreviata* Harper & Brenchley, 1993; *S.? globosa* (Williams, 1949), *S.? pulvinata* (Salter, 1864), *S. triangularis* (J. de C. Sowerby, 1839) and *S. turgida* (M'Coy, 1851). A number of unassigned species have been described from the British Isles by Williams (1963, p. 423), Lockley (1980, p. 210), Williams (1974, p. 114) and Parkes (1994, p. 152). None of these species are remotely similar to *S. gracilis* most of them being more coarsely costellate and showing various degrees of different external and internal morphologies.

Occurrence. – Tramore Limestone Formation, Units 2–3 and 5, Barrel Strand area. Dunabrattin Limestone Formation, Units I–III, Dunabrattin–Bunmahon area.

Order Pentamerida Schuchert & Cooper, 1931

Suborder Syntrophiidina Ulrich & Cooper, 1936

Superfamily Syntrophioidea Ulrich & Cooper, 1936

Family Porambonitidae Ulrich & Cooper, 1936

Genus *Hibernobonites* n. gen.

Type species (by original designation). – *Atrypa filosa* M'Coy, 1846 from the Upper Ordovician (Sandbian) slates of Knockmahon, Tramore, Co. Waterford.

Name. – From 'Hibernia', Latin for Ireland.

Diagnosis. – Ventral valve with relatively long, sub-parallel dental plates lacking a sessile spondylium; dorsal valve has massive brachiophores supported by short, divergent supporting plates, flanking a simple cardinal process. Interareas short to obsolete; shells rudimentary strophic. Ornament of radially arranged pits defined by fine fila and flat-topped bifurcating costellae.

Discussion. – The new genus is comparable to *Eoporambonites* Popov *et al.* 2005, and *Porambonites* Pander, 1830 but differs from both in features of its ornament and the rudimentary strophic condition of its shells. *Eoporambonites* is dorsibiconvex, multi-costellate with filae and lacks a pitted ornament; it also possesses relatively long brachiophore supporting plates. *Porambonites* is biconvex, costellate with lamellae and a pitted ornament together with relatively long brachiophore supporting plates. Popov *et al.* (2005) redefined the genus *Porambonites* to include only astrophic, smooth shells like the type species and they noted that the strophic shells of *Noetlingia* make affinity with the Porambonitidae doubtful. They suggested that *Noetlingia* could belong to the family Tetralobulidae and that it might be a senior synonym of *Punctolira* due to its pitted ornament. Zuykov *et al.* (2011) revised *Noetlingia* and placed it within the family Tetralobulidae with *Punctolira* as a junior synonym of *Noetlingia*. *Hibernobonites filosus* from Tramore and Dunabrattin, southeast Ireland, possesses a pitted surface like *Noetlingia* but is more aligned with the Porambonitidae rather than the Tetralobulidae; the latter lack a cardinal process and generally have a sessile spondylium. According to Popov *et al.* (2005), most of the Mid and Late Ordovician taxa presently assigned to *Porambonites* are not congeneric with the type species of the genus and probably represent new, undescribed, genera. Moreover, the morphology of most of the taxa presently assigned to *Porambonites* is only superficially known, many of the types are lost and details of their intrageneric morphological variation and phylogenetic relationships within the group unclear.

The Tramore form is assigned to '*Porambonites*' group, *sensu lato*, conforming to many of the characters defined for *Porambonites* by Carlson (2002), who included both smooth and pitted forms as well as strophic and astrophic shells. It is, however, doubtful that its shells are truly strophic and, thus, belong in the Porambonitidae, *sensu stricto* (see Popov *et al.* 2005). Cocks (2008) listed four formally described species of

'*Porambonites*' and four informally assigned forms including some with pitted ornament and shells like the Tramore form treated here. These are discussed further below.

Hibernobonites filosus (M'Coy, 1846)

Plate 9, figures 10–17; Text-figures 16, 17

1846 *Atrypa filosa* M'Coy, p. 39, pl. 3, fig. 28.

1869 *Porambonites intercedens* Pander; Davidson, p. 195 *pars*, pl. 26, fig. 3, *non* pl. 25, figs 17–19.

1869 *Porambonites intercedens* Pander var. *filosa* (M'Coy); Davidson, p. 195, pl. 25, fig. 16, pl. 26, figs 1, 2.

1978 *Porambonites filosus* (M'Coy); Cocks, p. 137.

2008 *Porambonites filosus* (M'Coy); Cocks, p. 173, pl. 7, fig. 10.

Remarks on the synonymy list. – The specimens illustrated by Davidson (1869, pl. 25, figs 17–19) were assigned to *Porambonites*? *maccoyanus* (Davidson, 1883) by Cocks (1978). The lectotype of *P.*? *maccoyanus* was selected by Cocks (1978) and figured by Cocks (2008, pl. 7, figs 2–4).

Material. – Total of seven ventral valves and five dorsal valves.

Diagnosis. – Same as genus.

Description. – Subelliptical to subpentagonal, strongly biconvex to dorsibiconvex *Hibernobonites* species with narrow, straight hingeline, 39% as wide as the medially situated maximum width (37–40%, $n = 2$); length 26–36 mm ($n = 9$) and width 32–43 mm ($n = 7$); ventral valve 78% as long as wide (68–87%, $n = 5$); posterolateral margins moderately strongly curved, lateral margins gently curved and anterior margin almost straight although it may be indented by a wide, shallow sulcus; sulcus poorly defined medially but becomes wider and more distinct anteriorly where it develops into a dorsally projecting tongue at the anterior margin, 52% as wide as the maximum valve width (50–54%, $n = 5$); dorsal valve 83% as long as wide (82–85%, $n = 2$); low dorsal fold poorly defined medially but becomes wider and more distinct anteriorly and gently folds the anterior margin, 40% as wide as the maximum valve width ($n = 1$); ventral and dorsal interareas short to obsolete, strongly apsacline and anacline respectively and may show prominent parallel growth lines; delthyrium and notothyrium open; delicate external ornament having the appearance of small, radially arranged pits defined by fine fila and flat-topped bifurcating costellae; seven costellae per mm at 5 mm anterior to the umbo; occasional strong concentric growth lines occurring more frequently at the margin.

Ventral interior (Fig. 54) with large, stout, peglike teeth; dental plates strong, thickened and subparallel but slightly converging anteriorly, 37% as long as the valve length (35–39%, $n = 3$); the dental plates enclose the diductor scars laterally and join with low, rounded ridges anteriorly, which define the anterior margin of the diductor scars and enclose the adductor scar; diductor scar well impressed, elongately oval, 58% as wide as long (54–61%, $n = 3$) and 32% as long as the valve length (30–35%, $n = 3$), composed of two slightly divergent scars separated by a low ridge that widens anteriorly; adductor scar subquadrate, well defined by the rounded ridges, almost as wide as the diductor scars and located just anterior of these, 14% as long as the valve length (13–15%, $n = 4$); mantle canal system saccate; *vascula myaria* consists of a large number of subparallel branches extending and slightly diverging anteriorly from around the adductor scar; radial ovarian ridges are impressed laterally and anterolaterally.

Dorsal interior (Fig. 55) with short, ridgelike cardinal process in the posterior end of the notothyrial cavity, 37% as long as the brachiophores (35–40%, $n = 2$); brachiophores stout, divergent, thickened in adult specimens, 71% as long as their lateral spread (61–77%, $n = 3$) and 22% as long as the valve length ($n = 1$); receding brachiophore bases join with low,

subparallel, rounded ridges extending for 34% the length of the valve (33–35%, $n = 2$) before converging abruptly to give a subrectangular extension to the notothyrial cavity; four well impressed, suboval adductor scars radiate from the anterolateral and anterior margins of the notothyrial cavity, separated by the median callist and a pair of thick, rounded lateral ridges extending from this; posterior adductors larger and more widely separated than the anterior adductors, total adductor scar 52% as long as wide (47–59%, $n = 3$) and 23% as long as the valve length ($n = 1$); mantle canal system poorly preserved; straight branches of the vascula media are observed radiating from between the anterior adductor scars as well as from the anterior margins of these.

Discussion. – *Hibernobonites filosus* from Tramore was first described by M'Coy (1846) as *Atrypa filosa* from slates of Sandbian age (Caradoc in Cocks 2008), Knockmahon, Co. Waterford. M'Coy's specimens may have been collected from the Dunabrattin Limestone Formation as Knockmahon is located about 3–5 km west of the gradational transition between the Tramore Limestone in the east and the Dunabrattin Limestone in the west (Fig. 2). Davidson (1869) collected his specimens from localities at Knockmahon, Dunabrattin and Tramore probably from both the Dunabrattin and Tramore Limestone formations like the material for this study. *Hibernobonites filosus* represents the largest brachiopod species recorded from both the Tramore and Dunabrattin Limestone Formation and is easily distinguished from the other species due to its characteristic morphology. M'Coy briefly described and

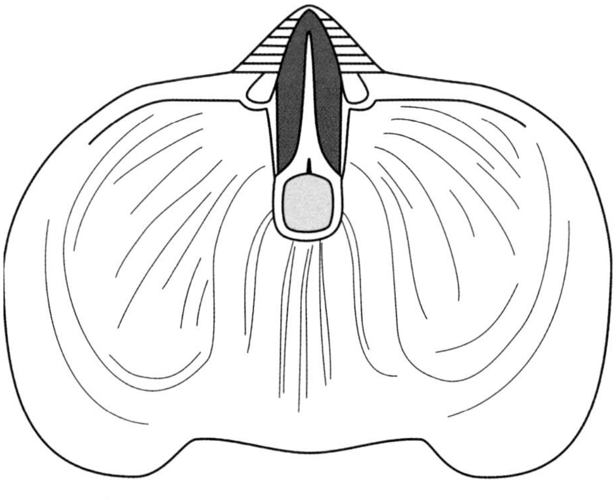

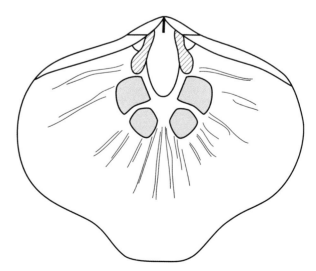

Fig. 54. Schematic drawing of the ventral interior of *Hibernobonites filosus*. Diductor muscle scar (dark grey), adductor muscle scar (light grey).

Fig. 55. Schematic drawing of the dorsal interior of *Hibernobonites filosus*. Cardinal process (thick, black line), brachiophores (hatched), adductor scars (light grey).

discussed *H. filosus* but his figured specimen (pl. 3, fig. 28) was unfortunately unsuitable for measuring. Davidson (1869) described and discussed the species in greater detail and some of his figured specimens (pl. 25, figs 16, 16a, 16c, pl. 26, figs 1–3) could be measured. The measurements were used for calculating ratios for comparison with the specimens from the current study, and the results are listed in Table 24. Despite the sparse amount of data, the ratio means and ranges of Davidson's and the current specimens are rather similar supporting that they belong to the same species.

Since the genus *Porambonites* was erected by Pander (1830), the validity of the species assigned to this genus through time has been extensively discussed. Reed (1899) collected a large number of *H. filosus* specimens (*Porambonites intercedens* var. *filosus* (M'Coy, 1846) in his publication) from the lower portion of, possibly, the Dunabrattin Limestone at Dunabrattin Head (Tramore Limestone in Reed 1899). He noted that *Porambonites intercedens* Pander, 1830; from the upper Dapingian – lower Darriwilian 'Orthoceras Limestone' (Pander 1830; pl. 11, figs 2a–e) was very similar to his specimens and that the Irish form (including the type specimen of *H. filosus*) possibly was a separate species. Pander originally described 30 species of '*Porambonites*' *sensu lato* from Northeast Europe but in a review of the family, Teichert (1930, p. 181) commented that seven of these did not belong to the genus, and of the remaining 23, only six, including *P. intercedens*, had been mentioned in the literature. He also noted that Pander's names only in rare cases were applicable as his diagnoses were very short and none of his original specimens were available for identification anymore. Teichert (1930) described 20 species of '*Porambonites*' *sensu lato* from the East Baltic provinces, but like Pander, his diagnoses tend to be brief and important internal features are not discussed or figured. Teichert revised some of Pander's species, including

P. intercedens (p. 186), which he referred to as a group rather than a species. He included four of Pander's original species *P. parallelum*, *P. surrecta*, *P. rotundata* and *P. rotunda* in this taxon and confined it to the middle Darriwilian (Kunda of the East Baltic. Like the Tramore form, the specimens of the *P. intercedens* group are distinguished by their weakly developed ornament; their smaller size compared to later forms; their subrounded shape, whereas younger forms have a more subtriangular to subpentagonal shape with an associated shorter hingeline; and by their shallow ventral sulcus and dorsal fold. Teichert mentioned the Dunabrattin/Tramore form (op. cit. p. 217) in his review of the distribution of the genus *Porambonites* in the Sandbian–early Katian and, like Reed (1899), he also suggested that the form described as *P. intercedens* var. *filosus* possibly was a distinct species but he did not discuss it further. Some of Pander's specimens originally described as *P. intercedens* were assigned to *P. filosus* and *P.? maccoyanus*, respectively, by Cocks (1978) and this study confirms the assignment to *H. filosus*.

Hibernobonites filosus was further compared to the British and Irish Darriwilian –Sandbian species presently assigned to '*Porambonites*' (see Cocks 2008), but only two are found similar enough for detailed comparison here:

Porambonites dubius Williams & Curry, 1985, was described from the Tourmakeady Limestone Formation, upper Floian – middle Dapingian (upper TS.2c–3a), Co. Mayo, west Ireland. It resembles *H. filosus* in having similar ventral and dorsal valve length:width ratios of 83% and 82%, respectively, a short anacline ventral and apsacline dorsal interarea as well as an open delthyrium and notothyrium. *Porambonites dubius* can be distinguished from *H. filosus*, however, by its smaller size, subelliptical to subtriangular outline, broad rectangular sulcus, longer dental plates, which are thickened anteriorly, and small triangular teeth. It has the same pitted

Table 24. Selected morphological ratios of *Hibernobonites filosus*.

Ratio	Valve	Data source	*n*	Mean %	Range %
*L*_valve × 100/*W*_valve	Ventral	Davidson (1869)	3	86	76–91
		This study	5	78	68–87
	Dorsal	Davidson (1869)	2	89	85–93
		This study	2	83	82–85
*W*_msc × 100/*L*_msc	Dorsal	Davidson (1869)	1	59	–
		This study	3	52	47–59
*W*_sulc × 100/*W*_valve	Ventral	Davidson (1869)	4	47	44–54
		This study	5	52	50–54
*W*_fold × 100/*W*_valve	Dorsal	Davidson (1869)	2	44	40–54
		This study	1	40	–

Ratios are from current investigation and measured from Davidson (1869, pl. 25, figs 16, 16a, 16c, pl. 26, figs 1–3).

ornamentation as *H. filosus*, although it is faint and much finer. We tentatively include the Tourmakeady species in our new genus.

Parkes (1994, p. 169) described *Porambonites* sp. from the Tramore Volcanic Formation, upper Sandbian (lowermost *TS*.5b–middle *TS*.5c), Kilbride, Co. Waterford, southeast Ireland. He noted that the sulcus of *Porambonites* sp. is similar to that of *H. filosus* but it can be distinguished from this by its smaller valve size, subtriangular outline and the longer dental plates, which extend for nearly 50% of the valve length. It too, however, can probably be included within *Hibernobonites*.

There are, possibly, a number of younger members of the genus from higher latitudes in the Upper Ordovician. For example, *Porambonites magnus* (Meneghini, 1880) from the Katian rocks of the Iberian Chains (Villas 1985) and *Porambonites (Porambonites) dreyfussl* Havlíček, 1981 from the Montagne Noire may also belong within *Hibernobonites*, the stock having migrated through time from low to high latitudes.

Occurrence. – Tramore Limestone Formation, Units 2–3, Barrel Strand area. Dunabrattin Limestone Formation, Units I–III, Dunabrattin–Bunmahon area.

Acknowledgements

The manuscript was part of the first author's PhD study and was financed by the Faculty of Science, University of Copenhagen for which the first author is very grateful. Flemming Groth-Hansen, Natural History Museum of Denmark, Copenhagen, Denmark, facilitated financial matters, which is deeply appreciated. Harper acknowledges support from the Leverhulme Trust and the Wenner-Gren Foundation. The first author sincerely thanks Sten L. Jacobsen, and M. Hofstedt, both at the Natural History Museum of Denmark, Copenhagen, Denmark, for producing latex and silicone casts of the Tramore brachiopods and Sten L. Jacobsen and B. W. Rasmussen, Natural History Museum of Denmark for technical support concerning photography. We are indebted to Matthew A. Parkes, National Museum of Ireland, Dublin, Ireland, for technical support in the initial stages of the research programme and to Yves Candela, National Museums Scotland, Edinburgh, UK, Lars Holmer, Uppsala University, Sweden, and Emma Sheldon, Geological Survey of Denmark and Greenland, Copenhagen, Denmark, for technical support in final stages of the manuscript. We gratefully acknowledge the help of Christian M. Ø. Rasmussen, Natural History Museum of Denmark for providing his Upper Ordovician brachiopod data, L. Robin M. Cocks, Natural History Museum, London, UK, in providing the brachiopod data of Lees et al. (2002) and R. B. Blodgett, consultant geologist, Anchorage, USA, Yves Candela, Jisuo Jin, Western University, London, Ontario, Canada, Ian G. Percival, Geological Survey of New South Wales, Australia, Jiayu Rong, State Key Laboratory of Palaeobiology and Stratigraphy, Nanjing, China, E. Villas, Universidad de Zaragoza, Spain, and Renbin Zhan, State Key Laboratory of Palaeobiology and Stratigraphy, Nanjing, China, for supplying additional data for the database. We thank the two reviewers, Drs Yves Candela and Leonid Popov, National Museum of Wales, Cardiff, UK, together with the editor Prof. Svend Stouge for their helpful comments. Financial support for the publication of this issue of Fossils and Strata was provided by the Lethaia Foundation. This is a contribution to IGCP 653, 'The onset of the Great Ordovician Biodiversification Event'.

References

Alikhova, T.N. 1951: *Brachiopods of the Middle and Upper parts of the Early Silurian of the Leningrad Region and their Stratigraphic Significance*, 81 pp. Vsesoyuznyj Nauchno-Issledovatel'skij Geologicheskij Institut (VSEGEI), Gosgeolizdat, Moscow.

Astini, R.A. 2003: The Ordovician proto-Andean basins. *In* Benedetto, J.L. (ed.): *Fossils of Argentina*, 1–74. Secretaría de Ciencia y Tecnología, Universidad Nacional de Córdoba, Córdoba.

Bachtadse, V. & Briden, J.C. 1990: Palaeomagnetic constraints on the position of Gondwana during Ordovician to Devonian times. *In* McKerrow, W.S. & Scotese, C.R. (eds): *Palaeozoic Palaeogeography and Biogeography. Geological Society, London, Memoirs, volume 12*, 43–48. The Geological Society of London, The Geological Society Publishing House, Bath.

Bancroft, B.B. 1945: The brachiopod zonal indices of the stages Costonian to Onnian in Britain. *Journal of Paleontology 19*, 181–252.

Barr, S.M. & White, C.E. 1996: Contrasts in late Precambrian-early Paleozoic tectonothermal history between Avalon composite terrane *sensu stricto* and other possible peri-Gondwanan terranes in southern New Brunswick and Cape Breton Island, Canada. *Geological Society of America Special Paper 304*, 95–108.

Bates, D.E.B. 1968: The lower Palaeozoic brachiopod and trilobite faunas of Anglesey. *Bulletin of the British Museum of Natural History (Geology) 16*, 125–199.

Benedetto, J.L. 2003a: Paleobiogeography. *In* Benedetto, J.L. (ed.): *Fossils of Argentina*, 91–109. Secretaría de Ciencia y Tecnología, Universidad Nacional de Córdoba, Córdoba.

Benedetto, J.L. 2003b: Brachiopods. *In* Benedetto, J.L. (ed.): *Fossils of Argentina*, 187–272. Secretaría de Ciencia y Tecnología, Universidad Nacional de Córdoba, Córdoba.

Bergström, S.M. & Orchard, M.J. 1985: Conodonts of the Cambrian and Ordovician systems from the British Isles. *In* Higgins, A.C. & Austin, A.L. (eds): *A Stratigraphical Index of Conodonts*, 32–67. Ellis Horwood, British Micropalaeontological Society, Chichester.

Bevins, R.E., Bluck, B.J., Brenchley, P.J., Fortey, R.A., Hughes, C.P., Ingham, J.K. & Rushton, A.W.A. 1992: Ordovician. *In* Cope, J.C.W., Ingham, J.K., Rawson, P.F. (eds): *Atlas of Palaeogeography and Lithofacies. Geological Society, London, Memoir, volume 13*, 19–36. The Geological Society of London, The Geological Society Publishing House, Bath.

Bluck, B.J., Gibbons, W. & Ingham, J.K. 1992: Terranes. *In* Cope, J.C.W., Ingham, J.K. & Rawson, P.F. (eds): *Atlas of Palaeogeography and Lithofacies. Geological Society, London, Memoirs, volume 13*, 1–4. The Geological Society, The Geological Society Publishing House, Bath.

Botquelen, A. & Mélou, M. 2007: Caradoc brachiopods from the Armorican Massif (northwestern France). *Journal of Paleontology 81*, 1080–1090.

Boucot, A.J. 1975: Evolution and extinction rate controls. *Developments in Palaeontology and Stratigraphy 1*, 427.

Boucot, A.J. 1993: Comments on Cambrian-to-Carboniferous biogeography and its implications for the Acadian Orogeny. *Geological Society of America Special Paper 275*, 41–49.

Brenchley, P.J. & Rawson, P.F. 2006: England and Wales through geological time. *In* Brenchley, P.J. & Rawson, P.F. (eds): *The Geology of England and Wales*, 1–7. Geological Society, London.

Brenchley, P.J., Mitchell, W.I. & Romano, M. 1977: A re-appraisal of some Ordovician successions in eastern Ireland. *Proceedings of the Royal Irish Academy 77B*, 65–85.

Brenchley, P.J., Rushton, A.W.A., Howells, M. & Cave, R. 2006: Cambrian and Ordovician: the early Palaeozoic tectonostratigraphic evolution of the Welsh Basin, Midland and Monian terranes of eastern Avalonia. *In* Brenchley, P.J. & Rawson, P.F.

(eds): *The Geology of England and Wales*, 25–74. Geological Society, London.

Brett, C.E., Boucot, A.J. & Jones, B. 1993: Absolute depths of Silurian benthic assemblages. *Lethaia 26*, 25–40.

Brück, P.M., Colthurst, J.R.J., Feely, M., Gardiner, P.R.R., Penney, S.R., Reeves, T.J., Shannon, P.M., Smith, D.G. & Vanguestaine, M. 1979: South-east Ireland: lower Palaeozoic stratigraphy and depositional history. *In* Harris, A.L., Holland, C.H. & Leake, B.E. (eds): *The Caledonides of the British Isles. Geological Society of London Special Publications, volume 8*, 533–544. Scottish Academic Press, Edinburgh.

Bruton, D.L. & Harper, D.A.T. 1981: Brachiopods and trilobites of the early Ordovician serpentine Otta Conglomerate, south central Norway. *Norsk Geologisk Tidsskrift 61*, 153–181.

Bruton, D.L. & Harper, D.A.T. 1985: Early Ordovician (Arenig–Llanvirn) faunas from oceanic islands in the Appalachian-Caledonian orogen. *In* Gee, D.G. & Sturt, B.A. (eds): *The Caledonian Orogen – Scandinavia and Related Areas*, 359–368. Wiley, Chichester.

Bryan, S.E., Cook, A.G., Evans, J.P., Hebden, K., Hurrey, L., Colls, P., Jell, J.S., Weatherley, D. & Firn, J. 2012: Rapid, long-distance dispersal by pumice rafting. *PLoS One 7*, 1–13.

Butts, C. 1942: Geology of the Appalachian Valley in Virginia, part 2. *Bulletin of the Geological Survey of Virginia 52*, pls. 41–97.

Candela, Y. 2003: Late Ordovician brachiopods from the Bardahessiagh Formation of Pomeroy, Ireland. *Monograph of the Palaeontographical Society London 156*, 95 pp.

Candela, Y. & Hansen, T. 2010: Brachiopod associations from the Middle Ordovician of the Oslo region, Norway. *Palaeontology 53*, 833–867.

Candela, Y. & Harper, D.A.T. 2010: Late Ordovician (Katian) brachiopods from the Southern Uplands of Scotland: biogeographic patterns on the edge of Laurentia. *Earth and Environmental Science Transactions of the Royal Society of Edinburgh 100*, 253–274.

Candela, Y. & Harper, D.A.T. 2014: Synoptic revision of the Ordovician brachiopods of the Barr and Lower Ardmillan groups of the Girvan area, Scotland. *Earth and Environmental Science Transactions of the Royal Society of Edinburgh 105*, 61–69.

Carlisle, H. 1979: Ordovician stratigraphy of the Tramore area, County Waterford, with a revised Ordovician correlation for south-east Ireland. *In* Harris, A.L., Holland, C.H. & Leake, B.E. (eds): *The Caledonides of the British Isles – Reviewed. Geological Society of London Special Publications, volume 8*, 545–554. Scottish Academic Press, Edinburgh.

Carlson, S.J. 2002: Suborder Syntrophiidina. *In* Kaesler, R.L. (ed.): *Treatise on Invertebrate Paleontology. Part H Brachiopoda Revised, volume 4*, 929–960. Geological Society of America, Boulder Colorado and the University of Kansas, Lawrence.

Channel, J.E.T., McCabe, C., Torsvik, T.H., Trench, T.A. & Woodcock, N.H. 1992: Palaeozoic palaeomagnetic studies in the Welsh Basin – recent advances. *Geological Magazine 129*, 533–542.

Christiansen, J.L. & Stouge, S. 1999: Oceanic circulation as an element in palaeogeographical reconstructions: the Arenig (early Ordovician) as an example. *Terra Nova 11*, 73–78.

Cocks, L.R.M. 1978: A review of British Lower Palaeozoic brachiopods, including a synoptic revision of Davidson's monograph. *Monograph of the Palaeontographical Society London 131*, 256.

Cocks, L.R.M. 2005: Strophomenate brachiopods from the Late Ordovician Boda Limestone of Sweden: their systematics and implications for palaeogeography. *Journal of Systematic Palaeontology 3*, 243–282.

Cocks, L.R.M. 2008: A revised review of British Lower Palaeozoic brachiopods. *Monograph of the Palaeontographical Society London 161*, 276 pp.

Cocks, L.R.M. 2010: Caradoc strophomenoid and plectambonitoid brachiopods from Wales and the Welsh Borderland. *Palaeontology 53*, 1155–1200.

Cocks, L.R.M. & Fortey, R.A. 1982: Faunal evidence for oceanic separations in the Palaeozoic of Britain. *Journal of the Geological Society 139*, 465–478.

Cocks, L.R.M. & Lockley, M.G. 1981: Reassessment of the Ordovician brachiopods from the Budleigh Salterton Pebble Bed, Devon. *Bulletin of the British Museum of Natural History (Geology) 35*, 111–124.

Cocks, L.R.M. & McKerrow, W.S. 1993: A reassessment of the early Ordovician 'Celtic' brachiopod province. *Journal of the Geological Society 150*, 1039–1042.

Cocks, L.R.M. & Rong, J.-Y. 1989: Classification and review of the brachiopod superfamily Plectambonitacea. *Bulletin of the British Museum of Natural History (Geology) 45*, 77–163.

Cocks, L.R.M. & Rong, J.-Y. 2000: Order Strophomenida. *In* Kaesler, R.L. (ed.): *Treatise on Invertebrate Paleontology. Part H Brachiopoda Revised, volume 2*, 216–248. Geological Society of America, Boulder Colorado and the University of Kansas, Lawrence.

Cocks, L.R.M. & Torsvik, T.H. 2002: Earth geography from 500 to 400 million years ago: a faunal and palaeomagnetic review. *Journal of the Geological Society 159*, 631–644.

Cocks, L.R.M. & Torsvik, T.H. 2011: The Palaeozoic geography of Laurentia and western Laurussia: a stable craton with mobile margins. *Earth-Science Reviews 106*, 1–51.

Cocks, L.R.M., McKerrow, W.S. & van Staal, C.R. 1997: The margins of Avalonia. *Geological Magazine 134*, 627–636.

Collins, A.S. & Buchan, C. 2004: Provenance and age constraints of the South Stack Group, Anglesey, UK: U-Pb SIMS detrital zircon data. *Journal of the Geological Society 161*, 743–746.

Colmenar, J. 2015: The arrival of brachiopods of the Nicolella Community to the Mediterranean margin of Gondwana during the Late Ordovician: palaeogeographical and palaeoecological implications. *Palaeogeography, Palaeoclimatology, Palaeoecology 428*, 12–20.

Conrad, T.A. 1843: Observations of the lead-bearing limestone of Wisconsin with descriptions of a new genus of trilobites and fifteen new Silurian fossils. *Journal of the Academy of Natural Sciences of Philadelphia 9*, 329–335.

Cooper, G.A. 1956: Chazyan and related brachiopods. *Smithsonian Miscellaneous Collections 127*, parts 1 and 2, 1245 pp.

Cooper, R.A. & Sadler, P.M. 2012: The Ordovician period. *In* Gradstein, F.M., Ogg, J.G., Schmitz, M. & Ogg, G. (eds): *The Geologic Time Scale, volume 2012*, 489–523. Elsevier B.V., Oxford, Amsterdam, Waltham.

Copper, P. 1977: *Zygospira* and some related Ordovician and Silurian atrypoid brachiopods. *Palaeontology 20*, 295–335.

Davidson, T. 1869: A monograph of the British Brachiopoda, Part VII, No. III. The Silurian Brachiopoda. *Monograph of the Palaeontographical Society London volume III*, 169–248.

Davidson, T. 1871: A monograph of the British Brachiopoda, Part VII, No. IV. The Silurian Brachiopoda. *Monograph of the Palaeontographical Society London volume III*, 249–397.

Davidson, T. 1883: A monograph of the British Brachiopoda, Part II. Silurian Supplement. *Monograph of the Palaeontographical Society London volume V*, 135–242.

Dewing, K. 1999: Late Ordovician and Early Silurian strophomenid brachiopods of Anticosti Island, Québec, Canada. *Palaeontographica Canadiana 17*, 143 pp.

Dewing, K. 2004: Shell structure and its bearing on the phylogeny of Late Ordovician–Early Silurian strophomenoid brachiopods from Anticosti Island, Quebec. *Journal of Paleontology 78*, 275–286.

Donovan, S.K. 1985: A cladid inadunate crinoid from the Ordovician Tramore Limestone of County Waterford. *Irish Journal of Earth Sciences 7*, 9–13.

Egerquist, E. 2003: New brachiopods from the Lower–Middle Ordovician (Billingen–Volkhov stages) of the East Baltic. *Acta Palaeontologica Polonica 48*, 31–38.

Fitton, J.G. & Hughes, D.J. 1970: Volcanism and plate tectonics in the British Ordovician. *Earth and Planetary Science Letters 8*, 223–228.

Fortey, R.A. & Cocks, L.R.M. 1988: Arenig to Llandovery faunal distributions in the Caledonides. *In* Harris, A.L. & Fettes, D.J. (eds): *The Caledonian-Appalachian Orogen. Geological Society of London Special Publications, volume 38*, 233–246. Scottish Academic Press, Edinburgh.

Fortey, R.A. & Cocks, L.R.M. 2003: Palaeontological evidence bearing on global Ordovician-Silurian continental reconstructions. *Earth-Science Reviews 61*, 245–307.

Fortey, R.A., Harper, D.A.T., Ingham, J.K., Owen, A.W. & Rushton, A.W.A. 1995: A revision of Ordovician series and stages from the historical type area. *Geological Magazine 132*, 15–30.

Fortey, R.A., Harper, D.A.T., Ingham, J.K., Owen, A.W., Parkes, M.A., Rushton, A.W.A. & Woodcock, N.H. 2000: A revised correlation of Ordovician rocks in the British isles. *Geological Society of London Special Reports 24*, 90 pp.

Freeman, G. & Lundelius, J.W. 2005: The transition from planktotrophy to lecithotrophy in larvae of Lower Palaeozoic Rhynchonelliform brachiopods. *Lethaia 38*, 219–254.

Fyffe, L.R., Barr, S.M., Johnson, S.C., McLeod, M.J., McNicoll, V.J., Valverde-Vaquero, P., van Staal, C.R. & White, C.E. 2009: Detrital zircon ages from Neoproterozoic and Early Paleozoic conglomerate and sandstone units of New Brunswick and coastal Maine: implications for the tectonic evolution of Ganderia. *Atlantic Geology 45*, 110–144.

Hall, J. 1847: Description of the organic remains of the lower division of the New-York System. *New York State Geological Survey, Palaeontology of New York 1*, 338 pp.

Hammer, Ø. & Harper, D.A.T. 2006: *Paleontological Data Analysis*, 351 pp. Blackwell Publishing, Malden, Oxford, Victoria.

Hammer, Ø., Harper, D.A.T. & Ryan, P.D. 2014: *PAST: Palaeontological Statistics software package for education and data analysis.* http://palaeo-electronica.org/2001_1/past/issue1_01.htm. Palaeontological Association.

Hansen, J. 2008: Upper Ordovician brachiopods from the Arnestad and Frognerkilen formations in the Oslo-Asker district, Norway. *Palaeontos 13*, 88 pp.

Hansen, J. & Harper, D.A.T. 2003: Brachiopod macrofaunal distribution through the upper Volkhov–lower Kunda (Lower Ordovician) rocks, Lynna River, St. Petersburg Region. *Bulletin of the Geological Society of Denmark 50*, 45–53.

Hansen, J. & Harper, D.A.T. 2005: *Palaeoneumania*, a new name for the genus *Neumania* Harper, 1981 (Brachiopoda), preoccupied by *Neumania* Lebert, 1879 (Arthropoda). *Norwegian Journal of Geology 85*, 223.

Hansen, J. & Holmer, L.E. 2010: Diversity fluctuations and biogeography of Ordovician brachiopod faunas in northeastern Spitsbergen. *Bulletin of Geosciences 85*, 497–504.

Hansen, J., Nielsen, J.K. & Hanken, N.-M. 2009: The relationships between Late Ordovician sea-level changes and faunal turnover in western Baltica: geochemical evidence of oxic and dysoxic bottom-water conditions. *Palaeogeography, Palaeoclimatology, Palaeoecology 271*, 268–278.

Harper, J.C. 1952: The Ordovician rocks between Collon (Co. Louth) and Grangegeeth (Co. Meath). *Scientific Proceedings Royal Dublin Society 26*, 85–112.

Harper, D.A.T. 1992: Ordovician provincial signals from Appalachian-Caledonian terranes. *Terra Nova 4*, 204–209.

Harper, D.A.T. 2000: Family Paurorthidae. *In* Kaesler, R.L. (ed.): *Treatise on Invertebrate Paleontology. Part H Brachiopoda Revised*, volume 3, 810 pp. Geological Society of America, Boulder Colorado and the University of Kansas, Lawrence.

Harper, D.A.T. 2006a: The Ordovician biodiversification: setting an agenda for marine life. *Palaeogeography, Palaeoclimatology, Palaeoecology 232*, 148–166.

Harper, D.A.T. 2006b: Brachiopods from the Upper Ardmillan Succession (Ordovician) of the Girvan district, Scotland. Part 3. *Monograph of the Palaeontographical Society London 159*, 129–187.

Harper, D.A.T. & Brenchley, P.J. 1993: An endemic brachiopod fauna from the Middle Ordovician of North Wales. *Geological Journal 28*, 21–36.

Harper, D.A.T. & Mac Niocaill, C. 2002: Early Ordovician rhynchonelliformean brachiopod biodiversity: comparing some platforms, margins and intra-oceanic sites around the Iapetus Ocean. *In* Crame, J.A. & Owen, A.W. (eds): *Palaeobiogeography and Biodiversity Change: The Ordovician and Mesozoic–Cenozoic Radiations. Geological Society of London Special Publications*, volume 194, 25–34. Scottish Academic Press, Edinburg.

Harper, D.A.T. & Parkes, M.E. 1989: Short paper: palaeontological constraints on the definition and development of Irish Caledonide terranes. *Journal of the Geological Society 146*, 413–415.

Harper, D.A.T. & Parkes, M.E. 2000: Ireland. *In* Fortey, R.A., Harper, D.A.T., Ingham, J.K., Owen, A.W., Parkes, M.A., Rushton, A.W.A. & Woodcock, N.H. (eds): *A Revised Correlation of the Ordovician Rocks of the British Isles. Geological Society of London Special Reports*, volume 24, 52–64. The Geological Society of London, The Geological Society Publishing House, Bath.

Harper, D.A.T., Mitchell, W.I., Owen, A.W. & Romano, M. 1985: Upper Ordovician brachiopods and trilobites from the Clashford House Formation, near Herbertstown, Co., Meath, Ireland. *Bulletin of the British Museum of Natural History (Geology) 38*, 287–308.

Harper, D.A.T., Mac Niocaill, C. & Williams, S.H. 1996: The palaeogeography of early Ordovician Iapetus terranes: an integration of faunal and palaeomagnetic constraints. *Palaeogeography, Palaeoclimatology, Palaeoecology 121*, 297–312.

Harper, D.A.T., Cocks, L.R.M., Popov, L.E., Sheehan, P.M., Bassett, M.G., Copper, P.M., Holmer, L.E., Jin, J. & Rong, J.-Y. 2004: Brachiopods. *In* Webby, B.D., Paris, F., Droser, M.L. & Percival, I.G. (eds): *The Great Ordovician Biodiversification Event*, 157–178. Columbia University Press, New York.

Harper, D.A.T., Owen, A.W. & Bruton, D.L. 2009: Ordovician life around the Celtic fringes: diversifications, extinctions and migrations of brachiopod and trilobite faunas at middle latitudes. *Geological Society of London Special Publications 325*, 157–170.

Harper, D.A.T., Parkes, M.A. & McConnell, B.J. 2010: Late Ordovician (Sandbian) brachiopods from the Mweelrea Formation, South Mayo, western Ireland: stratigraphic and tectonic implications. *Geological Journal 45*, 445–450.

Harper, D.A.T., Rasmussen, C.M.Ø., Liljeroth, M., Blodgett, R.B., Candela, Y., Jin, J., Percival, I.G., Rong, J.-Y., Villas, E. & Zhan, R.-B. 2013: Biodiversity, biogeography and phylogeography of Ordovician rhynchonelliform brachiopods. *In* Harper, D.A.T. & Servais, T. (eds): *Early Paleozoic Biogeography and Palaeogeography. Geological Society, London, Memoirs*, volume 38, 127–144. The Geological Society, The Geological Society Publishing House, Bath.

Harper, D.A.T., Parkes, M.A. & Ren-Bin, Z. 2017: Late Ordovician deep-water brachiopod faunas from Raheen, Waterford Harbour, Ireland. *Irish Journal of Earth Sciences 35*, 1–18.

Havlíček, V. 1952: On the Ordovician representatives of the family Plectambonitidae. *Sbornik Ústředního Ústavu geoloického 18*, 397–428.

Havlíček, V. 1967: Brachiopoda of the suborder Strophomenidina in Czechoslovakia. *Rozpravy Ústředního Ústavu geoloického 33*, 235.

Havlíček, V. 1970: Heterorthidae (Brachiopoda) in the Mediterranean Province. *Sborník Geologických věd paleontologie 12*, 7–41.

Havlíček, V. 1977: Brachiopods of the order Orthida in Czechoslovakia. *Rozpravy Ústředního Ústavu geologického 44*, 327.

Havlíček, V. 1981: Upper Ordovician Brachiopods from the Montagne Noire. *Palaeontographica Abteilung A 176*, 1–34.

Havlíček, V. & Vaněk, J. 1966: The biostratigraphy of the Ordovician of Bohemia. *Sbornik Geoloického věd Paleontologie 8*, 7–70.

Herrera, Z.A. & Benedetto, J.L. 1991: Early Ordovician brachiopod faunas of the Precordillera Basin, western Argentina: biostratigraphy and paleobiological affinities. *In* MacKinnon, D.I., Lee, D.E. & Campbell, J.D. (eds): *Brachiopods Through Time*, 283–301. Balkema, Rotterdam.

Hibbard, J.P., van Staal, C.R. & Miller, B.V. 2007: Links among Carolinia, Avalonia, and Ganderia in the Appalachian peri-Gondwanan realm. *In* Sears, J.W., Harms, T.A. & Evenchick, C.A. (eds): *Whence the Mountains? Inquiries into the Evolution of Orogenic Systems: A Volume in Honour of Raymond A. Price. The Geological Society of America Special Paper*, volume 433, 291–311. Geological Society of America, Boulder.

Hints, L. & Harper, D.A.T. 2003: Review of the Ordovician rhynchonelliformean Brachiopoda of the East Baltic: their distribution and biofacies. *Bulletin of the Geological Society of Denmark 50*, 29–43.

Hints, L. & Harper, D.A.T. 2008: The brachiopods *Alwynella* and *Grorudia*: homeomorphic plectambonitoids in the Middle and Upper Ordovician of Baltoscandia. *Earth and Environmental Science Transactions of the Royal Society of Edinburgh 98*, 271–280.

Hints, L. & Rõõmusoks, A. 1997: Ordovician articulate brachiopods. *In* Raukas, A. & Teedumäe, A. (eds): *Geology and Mineral Resources of Estonia*, 436. Estonian Academy Publishers, Tallinn.

Hofmann, H.J. 1963: Ordovician Chazy Group in southern Quebec. *American Association of Petroleum Geologists Bulletin 47*, 270–301.

Howe, H.J. 1988: Articulate brachiopods from the Richmondian of Tennessee. *Journal of Paleontology 62*, 204–218.

Huang R.X. (ed.) 2009: *Ocean Circulation. Wind-Driven and Thermohaline Processes*, 791 pp. Cambridge University Press, Cambridge.

Jaanusson, V. 1963: Lower and middle Viruan (middle Ordovician) of the Siljan District. *Bulletin of Geological Institutions of the University of Uppsala 42*, 1–40.

Jaanusson, V. 1973: Ordovician articulate brachiopods. *In* Hallam, A. (ed.): *Atlas of Palaeobiogeography*, 19–25. Elsevier, Amsterdam.

Jaanusson, V. 1976. Faunal dynamics in the Middle Ordovician (Viruan) of Baltoscandia. *In* Bassett, M.G. (ed.): *The Ordovician System. Proceedings of a Paleontological Association, Symposium, Birmingham, September 1974*, 301–326. University of Wales Press and National Museum of Wales, Cardiff.

Jaanusson, V. & Bassett, M.G. 1993: *Orthambonites* and related Ordovician brachiopod genera. *Palaeontology 36*, 21–63.

Jin, J. & Zhan, R.-B. 2000: Evolution of the Late Ordovician Orthid brachiopod *Gnamptorhynchos* Jin, 1989 from *Platystrophia* King, 1850 in North America. *Journal of Paleontology 74*, 983–991.

Jin, J., Caldwell, W.G.E. & Norford, B.S. 1997: Late Ordovician brachiopods and biostratigraphy of the Hudson Bay lowlands, northern Manitoba and Ontario. *Geological Survey of Canada Bulletin 513*, 115 pp.

Jones, O.T. 1928: *Plectambonites* and some allied genera. *Memoirs of the Geological Survey of Great Britain, Palaeontology 1*, 367–527.

Kaiser M.J., Attrill M.J., Jennings S., Thomas D.N., Barnes D.K.A., Brierley A.S., Polunin N.V.C., Raffaelli D.G., Williams P.J., Le B. (eds): 2005: *Marine Ecology. Processes, Systems, and Impacts*, 557 pp. Oxford University Press, New York.

Key, Jr, M.M., Jackson, P.N.W., Patterson, W.P. & Moore, M.D. 2005: Stable isotope evidence for diagenesis of the Ordovician Courtown and Tramore limestones, south-eastern Ireland. *Irish Journal of Earth Sciences 23*, 25–38.

Klinger, B.A. 1993: Gyre formation at a corner by rotating barotropic coastal flows along a slope. *Dynamics of Atmospheres and Oceans 19*, 27–63.

Kreisa, R.D. & Bambach, R.K. 1982: The role of storm processes generating shell beds in Paleozoic shelf environments. *In* Einsele, G. & Seilacher, A. (eds): *Cyclic and Event Stratification IIA*, 200–207. Springer, Berlin Heidelberg.

Lees, R.D.C., Fortey, R.A. & Cocks, L.R.M. 2002: Quantifying palaeogeography using biogeography: a test case for the Ordovician and Silurian of Avalonia based on brachiopods and trilobites. *Paleobiology 28*, 343–363.

Lockley, M.G. 1980: The Caradoc faunal associations of the area between Bala and Dinas Mawddwy, north Wales. *Bulletin of the British Museum of Natural History (Geology) 33*, 165–235.

Lockley, M.G. & Williams, A. 1981: Lower Ordovician Brachiopoda from mid and southwest Wales. *Bulletin of the British Museum of Natural History (Geology) 35*, 79.

Ludvigsen, R. 1975: Ordovician formations and faunas, Southern Mackenzie Mountains. *Canadian Journal of Earth Sciences 12*, 663–697.

Mac Niocaill, C., van der Pluijm, B.A. & Van der Voo, R. 1997: Ordovician paleogeography and the evolution of the Iapetus Ocean. *Geology 25*, 159–162.

MacGregor, A.R. 1961: Upper Llandeilo brachiopods from the Berwyn Hills, North Wales. *Palaeontology 4*, 177–209.

Maletz, J. & Egenhoff, S. 2011: Graptolite biostratigraphy and biogeography of the Table Head and Goose Tickle groups (Darriwilian, Ordovician) of western Newfoundland. *In* Guitérrez-Marco, J.C., Rábano, I. & García-Bellido, D. (eds): *Ordovician of the World*, 333–338. Instituto Geológico y Minero de España, Madrid.

Männil, R. 1966: *Evolution of the Baltic Basin during the Ordovician*. 200 pp. Valgus Publishers, Tallinn.

Maples, C.G. & Archer, A.W. 1988: Monte Carlo simulation of selected binomial similarity coefficients (II): effect of sparse data. *Palaios 3*, 9–103.

Max, M.D., Barber, A.J. & Martinez, J. 1990: Terrane assemblage of the Leinster Massif, SE Ireland, during the Lower Palaeozoic. *Journal of the Geological Society 147*, 1035–1050.

McConnell, B.J., Philcox, M.E., Sleeman, A.G., Stanley, G., Flegg, A.M., Daly, E.P. & Warren, W.P. 1994: *Geology of Kildare-Wicklow. A Geological Description to Accompany the Bedrock Geology 1:100,000 Map Series, Sheet 16, Kildare-Wicklow*. Geological Survey of Ireland, Dublin.

McNicoll, V.J., van Staal, C.R., Lentz, D. & Stern, R. 2002: Uranium-lead geochronology of Middle River rhyolite: implications for the provenance of basement rocks of the Bathurst mining camp, New Brunswick. *Radiogenic Age and Isotopic Studies: Report 15: Geological Survey of Canada Current Research 2002-F9*, 11 pp.

M'Coy, F. 1846: *A Synopsis of the Silurian Fossils of Ireland Collected from the Several Districts by Richard Griffith F.G.S.* 72 pp. Privately Published, Dublin.

M'Coy, F. 1851: On some new Cambro–Silurian fossils. *Annals and Magazine of Natural History (Series 2), 8*, 387–409.

Mélou, M. 1971: Nouvelle espèce de Leptestiina dans l'Ordovicien supérieur de l'aulne (Finistère). *Mémoires du Bureau de Recherches Géologiques et Minières 73*, 93–105.

Meneghini, G. 1880: Nuovi fossili siluriani di Sardegna. *Atti Reale Accademia Lincei, s. 3, Classe di Scienze, Fisiche, Matematiche, Naturali 5*, 209–219.

Mitchell, W.I. 1977: The Ordovician Brachiopoda from Pomeroy, Co. Tyrone. *Monograph of the Palaeontographical Society London 130*, 138 pp.

Mitchell, W.I., Carlisle, H., Hiller, N. & Addison, R. 1972: A correlation of the Ordovician rocks of Courtown (Co. Wexford) and Tramore (Co. Waterford). *Proceedings of the Royal Irish Academy (Section B) 72*, 25–38.

Möller, N.K. & Kvingan, K. 1988: The genesis of nodular limestones in the Ordovician and Silurian of the Oslo Region (Norway). *Sedimentology 35*, 405–420.

Muir-Wood, H. & Williams, A. 1965: Strophomenida. *In* Moore, R.C. (ed.): *Treatise on Invertebrate Paleontology. Part H Brachiopoda, volume 1*, H361–H521. Geological Society of America, Boulder Colorado and the University of Kansas, Lawrence.

Murphy, F.C., Anderson, T.B., Daly, J.S., Gallagher, V., Graham, J.R., Harper, D.A.T., Johnston, J.D., Kennan, P.S., Kennedy, M.J., Long, C.B., Morris, J.H., O'Keeffe, W.G., Parkes, M., Ryan, P.D., Sloan, R.J., Stillman, C.J., Tietzsch-Tyler, D., Todd, S.P. & Wrafter, J.P. 1991: An appraisal of Caledonian suspect terranes in Ireland. *Irish Journal of Earth Sciences 11*, 11–41.

Murphy, J.B., Pisarevsky, S.A., Nance, R.D. & Keppie, J.D. 2004: Neoproterozoic–early Paleozoic configuration of peri-Gondwanan terranes: implications for Laurentia-Gondwanan connections. *International Journal of Earth Sciences 93*, 659–682.

Neuman, R.B. 1964: Fossils in Ordovician tuffs, northeastern Maine. *U.S. Geological Survey Bulletin, 1181-E*, 38 pp.

Neuman, R.B. 1984: Geology and palaeobiology of islands in the Ordovician Iapetus Ocean: review and implications. *Geological Society of America Bulletin 95*, 1188–1201.

Neuman, R.B. & Bates, D.E.B. 1978: Reassessment of Arenig and Llanvirn age (Early Ordovician) brachiopods from Anglesey, north-west Wales. *Palaeontology 21*, 571–613.

Neuman, R.B. & Harper, D.A.T. 1992: Paleogeographic significance of Arenig-Llanvirn Toquima-Table Head and Celtic brachiopod assemblages. *In* Webby, B.D. & Laurie, J.R. (eds): *Global Perspectives on Ordovician Geology*, 241–254. Balkema, Rotterdam.

Nielsen, A.T. 2004: Ordovician sea level changes: a Baltoscandian perspective. *In* Webby, B.D., Paris, F., Droser, M.L. & Percival, I.G. (eds): *The Great Ordovician Biodiversification Event*, 84–93. Columbia University Press, New York, Chichester, West Sussex.

Nikitin, I.F., Popov, L.E. & Bassett, M.G. 2003: Late Ordovician brachiopods from the Selety river basin, North Central Kazakhstan. *Acta Palaeontologica Polonica* 48, 39–54.

Öpik, A. 1930: Brachiopoda, Protremata der Estländischen Ordovizischen Kukruse-stufe. *Acta et Commentationes Universitatis Tartuensis (Dorpatensis) (Series A)* 17, 162 pp.

Öpik, A. 1933a: Über Plectamboniten. *Acta et Commentationes Universitatis Tartuensis (Dorpatensis) (Series A)* 24, 79 pp.

Öpik, A. 1933b: Über einige Dalmanellacea aus Estland. *Acta et Commentationes Universitatis Tartuensis (Dorpatensis) (Series A)* 25, 25 pp.

Owen, A. & Parkes, M.A. 2000: Trilobite faunas of the Duncannon Group: Caradoc stratigraphy, environments and palaeobiogeography of the Leinster Terrane, Ireland. *Palaeontology* 43, 219–269.

Pander, C.H. 1830: *Beiträge zur Geognosie des Russischen reiches*, 165 pp. K. Kray, St. Petersburg.

Parkes, M.A. 1992: Caradoc brachiopods from the Leinster terrane (SE Ireland) – a lost piece of the Iapetus puzzle? *Terra Nova* 4, 223–230.

Parkes, M.A. 1994: The brachiopods of the Duncannon Group (Middle–Upper Ordovician) of south-east Ireland. *Bulletin of the British Museum of Natural History (Geology)* 50, 105–174.

Parkes, M.A. & Harper, D.A.T. 1996: Ordovician brachiopod biogeography in the Iapetus suture zone of Ireland: provincial dynamics in a changing ocean. *In* Copper, P. & Jin, J. (eds): *Brachiopods*, 197–202. Balkema, Rotterdam.

Phillips, W.E.A., Stillman, C.J. & Murphy, T. 1976: A Caledonian plate tectonic model. *Journal of the Geological Society* 132, 579–609.

Pohl, A., Donnadieu, Y., Buoncristiani, J.-F. & Vennin, E. 2014: Effect of the Ordovician paleogeography on the (in)stability of the climate. *Climate of the Past* 10, 2053–2066.

Pohl, A., Nardin, E., Vandenbroucke, T.R.A. & Donnadieu, Y. 2016a: High dependence of Ordovician ocean surface circulation on atmospheric CO^2 levels. *Palaeogeography, Palaeoclimatology, Palaeoecology* 458, 39–51.

Pohl, A., Donnadieu, Y., Le Hir, G., Ladant, J.-B., Dumas, C., Alvarez-Solas, J. & Vandenbroucke, T.R.A. 2016b: Glacial onset predated Late Ordovician climate cooling. *Paleoceanography* 31, 800–821.

Pollock, J.C., Hibbard, J.P. & van Staal, C.R. 2012: A paleogeographical review of the peri-Gondwanan realm of the Appalachian orogeny. *Canadian Journal of Earth Sciences* 49, 259–288.

Popov, L.E., Egerquist, E. & Zuykov, M.A. 2005: Ordovician (Arenig–Caradoc) syntrophiidine brachiopods from the East Baltic region. *Palaeontology* 48, 739–761.

Portlock, J.E. 1843: *Report of the Geology of the County of Londonderry and of Parts of Tyrone and Fermanagh.* 784 pp. Andrew Milliken, Dublin.

Pothier, H.D., Waldron, J.W.F., Schofield, D.I. & DuFrane, S.A. 2015: Peri-Gondwanan terrane interactions recorded in the Cambrian–Ordovician detrital zircon geochronology of North Wales. *Gondwana Research* volume 28, 987–1001.

Potter, A.W. & Boucot, A.J. 1992: Middle and Late Ordovician brachiopod benthic assemblages of North America. *In* Webby, B.D. & Laurie, J.R. (eds): *Global Perspectives on Ordovician Geology*, 307–323. Balkema, Rotterdam.

Rasmussen, J.A. 1998: A reinterpretation of the conodont Atlantic realm in the late Early Ordovician (Early Llanvirn). *Palaeontologia Polonica* 58, 67–77.

Rasmussen, C.M.Ø. 2005: Middle Ordovician (Kundan) brachiopods from the East Baltic: aspects of taxonomy, biogeography, palaeoecology, bio- and ecostratigraphy. Cand. Scient. Thesis, University of Copenhagen (unpublished).

Rasmussen, C.M.Ø., Hansen, J. & Harper, D.A.T. 2007: Baltica: a mid Ordovician diversity hotspot. *Historical Biology* 19, 255–261.

Rasmussen, C.M.Ø. & Harper, D.A.T. 2008: Resolving early Mid Ordovician (Kundan) bioevents in the East Baltic based on brachiopods. *Geobios* 41, 533–542.

Rasmussen, C.M.Ø., Harper, D.A.T. & Blodgett, R.B. 2012: Late Ordovician brachiopods from West-central Alaska: systematics, ecology and palaeobiogeography. *Fossils and Strata* 58, 103 pp.

Rasmussen, C.M.Ø., Ullmann, C.V., Jakobsen, K.G., Lindskog, A., Hansen, J., Hansen, T., Eriksson, M.E., Dronov, A., Frei, R., Korte, C., Nielsen, A.T. & Harper, D.A.T. 2016: Onset of main Phanerozoic marine radiation sparked by emerging Mid Ordovician icehouse. *Scientific Reports* 6, 1–9.

Reed, F.R.C. 1899: The lower Palaeozoic bedded rocks of Co., Waterford. *Quarterly Journal of the Geological Society* 55, 718–772.

Reed, F.R.C. 1917: The Ordovician and Silurian Brachiopoda of the Girvan District. *Transactions of the Royal Society of Edinburgh* 51, 116–998.

Richardson, J.R. 1997: Biogeography of articulated brachiopods. *In* Kaesler, R.L. (ed.): *Treatise on Invertebrate Paleontology. Part H Brachiopoda Revised, volume 1*, 461–472. Geological Society of America, Boulder Colorado and the University of Kansas, Lawrence.

Rõõmusoks, A. 2004: Ordovician strophomenoid brachiopods of northern Estonia. *Fossilia Baltica* 3, 151 pp.

Ross, Jr, R.J. & Ingham, J.K. 1970: Distribution of the Toquima-Table Head (Middle Ordovician Whiterock) faunal realm in the Northern Hemisphere. *Geological Society of America Bulletin* 81, 393–408.

Rubel, M. 1961: Lower Ordovician brachiopods of the superfamilies Orthacea, Dalmanellacea and Syntrophiacea of Eastern Baltic. *ENSV Teaduste Academia Geologia Instituudi Uurimused* 6, 141–226.

Salter, J.W. 1849: Notes on the fossils from the limestone on the Stincher River, and from the slates of Loch Ryan. *Quarterly Journal of the Geological Society* 5, 13–17.

Salter, J.W. 1864: Notes on the fossils from the Budleigh Salterton Pebble-bed. *Quarterly Journal of the Geological Society* 20, 286–302.

Schuchert, C. & Cooper, G.A. 1932: Brachiopod genera of the suborder Orthoidea and Pentameroidea. *Memoirs of the Peabody Museum of Natural History* 41, 270 pp.

Sedgwick, A. & M'Coy, F. 1851–55: *A Synopsis of the Classification of the British Palaeozoic Rocks, with a Systematic Description of the British Palaeozoic Fossils in the Geological Museum of the University of Cambridge*, xcviii + 661 pp. London: John W. Parker and son, West Strand, Cambridge: Deighton, Bell & CO. M.DCCC.LV.

Sepkoski, Jr, J.J. & Sheehan, P.M. 1983: Diversification, faunal change, and community replacement during the Ordovician radiations. *Topics in Paleobiology* 3, 673–717.

Servais, T., Owen, A.W., Harper, D.A.T., Kröger, B. & Munnecke, A. 2010: The Great Ordovician Biodiversification Event (GOBE): the palaeoecological dimension. *Palaeogeography, Palaeoclimatology, Palaeoecology* 294, 99–119.

Servais, T., Danelian, T., Harper, D.A.T. & Munnecke, A. 2014: Possible oceanic circulation patterns, surface water currents and upwelling zones in the Early Palaeozoic. *GFF* 136, 229–233.

Sleeman, A.G. & McConnell, B. 1995: *Bedrock Geology 1:100,000 Map Series, East CORK – Waterford – Sheet 22.* Geological Survey of Ireland, Dublin.

Sowerby, J.de.C. 1839: Shells. *In* Murchison, R.I. (ed.): *The Silurian System*, 579–712. John Murray, London.

Spjeldnæs, N. 1957: The Middle Ordovician of the Oslo Region, Norway. 8. Brachiopods of the suborder Strophomenida. *Norsk Geologisk Tidsskrift* 37, 214 pp.

van Staal, C.R., Sullivan, R.W. & Whalen, J.B. 1996: Provenance and tectonic history of the Gander Zone in the Caledonian/

Appalachian orogeny: implications for the origin and assembly of Avalon. *Geological Society of America Special Papers 304*, 347–367.

van Staal, C.R., Dewey, J.F., Mac Niocaill, C. & McKerrow, W.S. 1998: The Cambrian–Silurian tectonic evolution of the northern Appalachians and British Caledonides: history of a complex, west and southwest Pacific-type segment of Iapetus. *In* Blundell, D.J. & Scott, A.C. (eds): *Lyell: The Past is the Key to the Present. Geological Society of London Special Publications*, volume 143, 199–242. The Geological Society of London, The Geological Society Publishing House, Bath.

van Staal, C.R., Whalen, J.B., Valverde-Vaquero, P., Zagorevski, A. & Rogers, N. 2009: Pre-Carboniferous, episodic accretion-related, orogenesis along the Laurentian margin of the Northern Appalachians. *Geological Society of London Special Publications 327*, 271–316.

van Staal, C.R., Barr, S.M. & Murphy, J.B. 2012: Provenance and tectonic evolution of Ganderia: constraints on the evolution of the Iapetus and Rheic oceans. *Geology 40*, 987–990.

Stillman, C.J. 1978: South-east County Waterford and South Tipperary: ordovician volcanic and sedimentary rocks and Silurian turbidites. *In* Brück, P.M. & Naylor, D. (eds): *Field Guide to the Caledonian and Pre-Caledonian Rocks of South-East Ireland. Geological Survey of Ireland, Guide Series*, volume 2, 41–60. Geological Survey of Ireland, Dublin.

Stillman, C.J. 1986: A comparison of the lower Palaeozoic volcanic rocks on either side of the Caledonian suture in the British Isles. *In* Fettes, D.J. & Harris, A.L. (eds): *Synthesis of the Caledonian rocks of Britain*, 187–205. Reidel Publishing, Dordrecht.

Teichert, C. 1930: Biostratigraphie der Poramboniten. Eine entwicklungsgeschichtliche, paläontologische und vergleichen-geotektonische studie. *Neues Jahrbuch für Mineralogie, Geologie und Paläontologie (Abhandlungen B) 63*, 177–246.

Tietzsch-Tyler, D. & Sleeman, A.G. 1994: *Geology of South Wexford. A Geological Description of South Wexford and Adjoining Parts of Waterford, Kilkenny and Carlow to Accompany the Bedrock Geology 1:100,000 Scale Map Series, Sheet 23, South Wexford*. Geological Survey of Ireland, Dublin.

Torsvik, T.H. 2014: *BugPlates: Linking biogeography and palaeogeography*. Geological Survey of Norway. http://www.geodynamics.no/bugs/SoftwareManual.pdf.

Torsvik, T.H. & Cocks, L.R.M. 2009: The Lower Palaeozoic palaeogeographical evolution of the northeastern and eastern peri-Gondwanan margin from Turkey to New Zealand. *In* Bassett, M.G. (ed.): *Early Palaeozoic Peri-Gondwana Terranes: New Insights fro"' Tectonics w1d Biogeography. Geological Society, London, Special Publications, 325*, 3–21. The Geological Society of London, The Geological Society Publishing House, Bath.

Torsvik, T.H. & Cocks, L.R.M. 2013: New global palaeogeographical reconstructions for the Early Palaeozoic and their generation. *In* Harper, D.A.T. & Servais, T. (eds): *Early Palaeozoic Biogeography and Palaeogeography. Geological Society, London, Memoirs, volume 38*, 5–24. The Geological Society, The Geological Society Publishing House, Bath.

Torsvik, T.H., Smethurst, M.A., Briden, J.C. & Sturt, B.A. 1990: A review of Palaeozoic palaeomagnetic data from Europe and their palaeogeographic implications. *In* McKerrow, W.S. & Scotese, C.R. (eds): *Palaeozoic Palaeogeography and Biogeography. Geological Society, London, Memoirs, volume 12*, 25–41. The Geological Society of London, The Geological Society Publishing House, Bath.

Torsvik, T.H., Smethurst, M.A., Van der Voo, R., Trench, A., Abrahamsen, N. & Halvorsen, E. 1992: Baltica. A synopsis of Vendian-Permian palaeomagnetic data and their palaeotectonic implications. *Earth-Science Reviews 33*, 133–152.

Torsvik, T.H., Meert, J.G., Van der Voo, R., McKerrow, W.S., Brasier, M.D., Sturt, B.A. & Walderhaug, H.J. 1996: Continental break-up and collision in the Neoproterozoic and Palaeozoic – a tale of Baltica and Laurentia. *Earth Science Reviews 40*, 229–258.

Trench, A. & Torsvik, T.H. 1992: Palaeomagnetic constraints on the Early-Middle Ordovician palaeogeography of Europe: recent advances. *In* Webby, B.D. & Laurie, J.R. (eds): *Global perspectives on Ordovician Geology*, 255–259. Balkema, Rotterdam.

Trench, A., Torsvik, T.H. & McKerrow, W.S. 1992: The palaeogeographic evolution of southern Britain during Early Palaeozoic times: a reconciliation of palaeomagnetic and biogeographic evidence. *Tectonophysics 201*, 75–82.

Tucker, M.E. 1991: *Sedimentary Petrology an Introduction to the Origin of Sedimentary Rocks*. 260 pp. Blackwell Science, University Press, Cambridge.

Ulrich, E.O. & Cooper, G.A. 1936: New genera and species of Ozarkian and Canadian brachiopods. *Journal of Paleontology 10*, 620–626.

Ulrich, E.O. & Cooper, G.A. 1942: New genera of Ordovician brachiopods. *Journal of Paleontology 16*, 616–631.

Valverde-Vaquero, P., van Staal, C.R., McNicoll, V. & Dunning, G.R. 2006: Mid-Late Ordovician magmatism and metamorphism along the Gander margin in central Newfoundland. *Journal of the Geological Society 163*, 347–362.

Van der Voo, R., Johnson, R.J.E., van der Pluijm, B.A. & Knutson, L.C. 1991: Paleogeography of some vestiges of Iapetus: paleomagnetism of the Ordovician Robert's Arm, Summerford, and Chanceport groups, central Newfoundland. *Geological Society of America Bulletin 103*, 1564–1575.

Villas, E. 1985: Braquiopodos del Ordovicico medio y superior de las Cadenas Ibericas Orientales. *Memorias del Museo Paleontologico de la Universidad de Zaragoza 1*, 223 pp.

Villas, E. 1992: New Caradoc brachiopods from the Iberian Chains (northeastern Spain) and their stratigraphic significance. *Journal of Paleontology 66*, 772–793.

Villas, E., Vizcaïno, D., Álvaro, J.J., Destombes, J. & Vennin, E. 2006: Biostratigraphic control of the latest Ordovician glaciogenic unconformity in Alnif (eastern Anti-Atlas, Morocco), based on brachiopods. *Geobios 39*, 727–737.

Villas, E., Vennin, E., Jiménez-Sánchez, A. & Zamora, S. 2011: Day 2. The Upper Ordovician formations of the eastern Iberian Chain. *11th International Symposium on the Ordovician System, post-symposium field trip guide*, 10–19.

Waldron, J.W.F., Schofield, D.I., Dufrane, S.A., Floyd, J.D., Crowley, Q.G., Simonetti, A., Dokken, R.J. & Pothier, H.D. 2014: Ganderia-Laurentia collision in the Caledonides of Great Britain and Ireland. *Journal of the Geological Society 171*, 555–569.

Wang, Y. 1949: Maquoketa Brachiopoda of Iowa. *Geological Society of America Memoir 42*, 1–52.

Webby, B.D., Ross, Jr, R.J. & Zhen, Y.Y. 1994: The Ordovician system of the East European platform and Tuva (southeastern Russia). *International Union of Geological Sciences 28*, 61 pp.

Webby, B.D., Cooper, R.A., Bergström, S.M. & Paris, F. 2004: Stratigraphic framework and time slices. *In* Webby, B.D., Paris, F., Droser, M.L. & Percival, I.G. (eds): *The Great Ordovician Biodiversification Event*, 41–47. Columbia University Press, New York.

Whittington, H.B. 1938: New Caradocian brachiopods from the Berwyn Hills, North Wales. *Annals and Magazine of Natural History 11*, 241–259.

Whittington, H.B. & Williams, A. 1955: The fauna of the Derfel Limestone of the Arenig District, North Wales. *Philosophical Transactions of the Royal Society B238*, 397–430.

Williams, A. 1956: Productorthis in Ireland. *In Proceedings of the Royal Irish Academy. Section B: Biological, Geological and Chemical science, volume 57*, 179–183. Royal Irish Academy, Dublin 1954–1956.

Williams, A. 1949: New Lower Ordovician brachiopods from the Llandeilo-Llangadock District. *Geological Magazine 86*, 226–238.

Williams, A. 1962: The Barr and Lower Ardmillan Series (Caradoc) of the Girvan district, South-west Ayrshire, with descriptions of the Brachiopoda. *Geological Society, London, Memoirs, volume 3*, 267 pp., The Geological Society of London, Surrey.

Williams, A. 1963: The Caradocian brachiopod faunas of the Bala District, Merionethshire. *Bulletin of the British Museum of Natural History (Geology)* 8, 327–471.

Williams, A. 1969: Ordovician of the British Isles. *American Association of Petroleum Geologists Memoir 12*, 236–264.

Williams, A. 1973: Distribution of brachiopod assemblages in relation to Ordovician palaeogeography. *Special Papers in Palaeontology 12*, 241–269.

Williams, A. 1974: Ordovician Brachiopoda from the Shelve district Shropshire. *Bulletin of the British Museum of Natural History (Geology) Supplement 11*, 163 pp.

Williams, A. 1976: Ireland. *In* Williams, A., Strachan, I., Bassett, D.A., Dean, W.T., Ingham, J.K., Wright, A.D. & Whittington, H.B. (eds): *A Correlation of Ordovician Rocks in the British Isles. Geological Society of London Special Reports, volume 3*, 53–74. Scottish Academic Press, Edinburgh.

Williams, A. & Brunton, C.H.C. 1997: Morphological anatomical terms applied to brachiopods. *In* Kaesler, R.L. (ed.): *Treatise on Invertebrate Paleontology. Part H Brachiopoda Revised, volume 1*, 421–440. Geological Society of America, Boulder Colorado and the University of Kansas, Lawrence.

Williams, A. & Curry, G.B. 1985: Lower Ordovician Brachiopoda from the Tourmakeady Limestone, Co., Mayo, Ireland. *Bulletin of the British Museum of Natural History (Geology) 38*, 183–269.

Williams, A. & Harper, D.A.T. 2000: Family Orthidae. *In* Kaesler, R.L. (ed.): *Treatise on Invertebrate Paleontology. Part H Brachiopoda Revised, volume 3*, 724–728. Geological Society of America, Boulder Colorado and the University of Kansas, Lawrence.

Williams, A. & Wright, A.D. 1963: The classification of the "*Orthis testudinaria* Dalman" group of brachiopods. *Journal of Paleontology 37*, 1–32.

Williams, A., Lockley, M.G. & Hurst, M. 1981: Benthic palaeocommunities represented in the Ffairfach Group and coeval Ordovician successions of Wales. *Palaeontology 24*, 661–694.

Williams, S.H., Harper, D.A.T., Neuman, R.B., Boyce, W.D. & Mac Niocaill, C. 1996: Lower Paleozoic fossils from Newfoundland and their importance in understanding the history of the Iapetus Ocean. *Geological Survey of Canada, Ottawa, Papers 42*, 115–126.

Wonderley, P.F. & Neuman, R.B. 1984: The Indian Bay Formation: fossiliferous Early Ordovician volcanogenic rocks in the northern Gander Terrane, Newfoundland. *Canadian Journal of Earth Sciences 21*, 525–532.

Woodcock, N.H. 2000: Terranes in the British and Irish Ordovician. *In* Fortey, R.A., Harper, D.A.T., Ingham, J.K., Owen, A.W., Parkes, M.A., Rushton, A.W.A. & Woodcock, N.H.

(eds): *A Revised Correlation of the Ordovician Rocks of the British Isles. Geological Society of London Special Reports, volume 24*, 8–12. The Geological Society of London, The Geological Society Publishing House, Bath.

Wright, A.D. 1964: The fauna of the Portrane Limestone II. *Bulletin of the British Museum of Natural History (Geology) 9*, 137–256.

Wright, A.D. & Stigall, A.L. 2014: Species-level phylogenetic revision of the Ordovician orthide brachiopod *Glyptorthis* from North America. *Journal of Systematic Palaeontology 12*, 893–906.

Wyse Jackson, P.N., Buttler, C.J. & Key, Jr, M.M. 2001: Palaeoenvironmental interpretation of the Tramore Limestone Formation (Llandeilo, Ordovician) based on bryozoan colony form. *In* Wyse Jackson, P.N., Buttler, C.J. & Spencer Jones, M.E. (eds): *Bryozoan Studies 2001*, 359–365. Balkema, Lisse.

Wysogórski, J. 1900: Zur entwicklungsgeschichte der Brachiopodenfamilie der Orthiden im ostbaltischen Silur. *Zeitschrift der Deutschen Geologischen Gesellschaft 52*, 222–236.

Xu, H.-K. & Liu, D.-Y. 1984: Late Lower Ordovician brachiopods of southwestern China. *Bulletin of Nanjing Institute of Geology and Palaeontology 8*, 147–235.

Zagorevski, A., van Staal, C.R., Rogers, N., McNicoll, V.J. & Pollock, J. 2010: Middle Cambrian to Ordovician arc-backarc development on the leading edge of Ganderia, Newfoundland Appalachians. *In* Tollo, R.P., Bartholomew, M.J., Hibbard, J.P. & Karabinos, M.P. (eds): *From Rodinia to Pangea: The Lithotectonic Record of the Appalachian Region. Geological Society of America Memoir, volume 206*, 367–396. The Geological Society of America, Boulder.

Zhan, R.-B. & Jin, J. 2005: Brachiopods from the Middle Ordovician Shihtzupu Formation of Yunnan Province, China. *Acta Palaeontologica Polonica 50*, 365–393.

Zuykov, M.A. & Harper, D.A.T. 2004: *Platystrophia* King, 1850 (Brachiopoda, Orthida): proposed conservation of usage by designation of *Porambonites costatus* Pander, 1830 (currently *Platystrophia costata*) as the type species of *Platystrophia*. *Bulletin of Zoological Nomenclature 61*, 246–250.

Zuykov, M. & Harper, D.A.T. 2007: *Platystrophia* (Orthida) and new related Ordovician and early Silurian brachiopod genera. *Estonian Journal of Earth Sciences 56*, 11–34.

Zuykov, M., Harper, D.A.T., Terentiev, S.S. & Pelletier, E. 2011: Revision of the plectorthoid brachiopod *Platystrophia dentata* (Pander, 1830) from the Middle Ordovician of the East Baltic. *Estonian Journal of Earth Sciences 60*, 131–136.

2008-2014 SARV: Geoscience collections and data repository. Geoscience Collections of Estonia: http://geokogud.info/search.php

Plates 1–10

Plate 1

1–11:   *Dactylogonia costellata* **n. sp.**
1   Interior mould of dorsal valve, NHMUK PI BB35218. Scale bar = 4 mm.
2   Latex cast of dorsal interior. Scale bar = 4 mm. Specimen lost.
3, 5   Interior mould and latex cast of dorsal valve, NHMUK PI BB35217. Scale bar = 5 mm.
4, 11   Exterior mould and lateral view of ventral valve, NHMUK PI BB35220. Scale bar = 3 mm.
6, 7   Holotype. Interior mould and latex cast of ventral valve, NHMUK PI BB35216. Scale bar = 5 mm.
8   Interior mould of ventral valve, NHMUK PI BB35221. Scale bar = 5 mm.
9, 10   Exterior mould and latex cast of ventral valve, NHMUK PI BB35219. Scale bar = 5 mm.

12:   *Tetraphalerella***? sp.** Latex cast of dorsal interior. Scale bar = 10 mm. Specimen lost.

13–16: *Colaptomena auduni* **n. sp.**
13, 14   Interior mould and latex cast of dorsal valve, NHMUK PI BB35208. Scale bar = 10 mm.
15, 16   Interior mould and latex cast of dorsal valve, NHMUK PI BB35209. Scale bar = 10 mm.

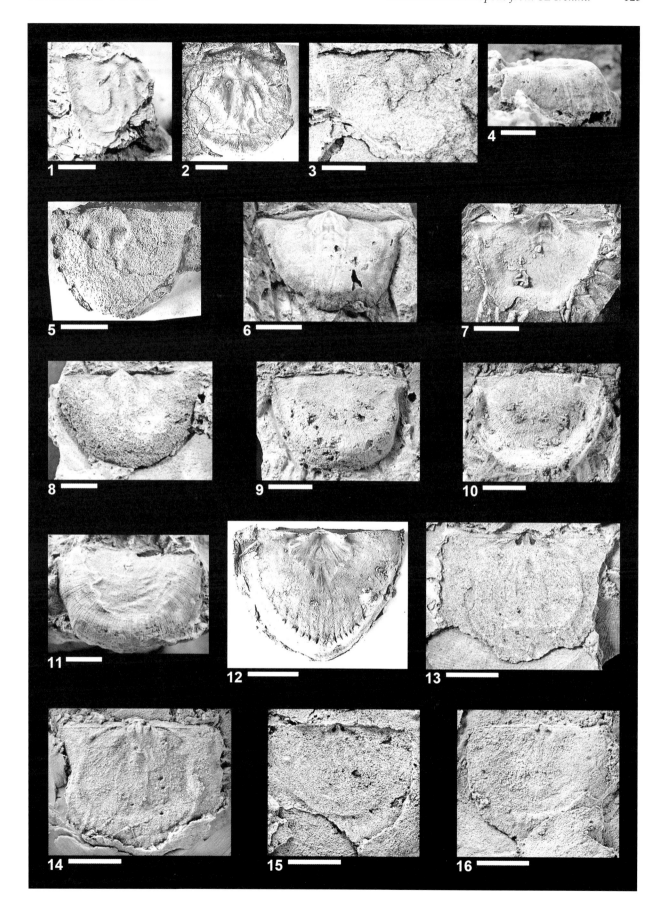

Plate 2

1–8:   *Colaptomena auduni* **n. sp.**

1, 2   Holotype. Interior mould and latex cast of ventral valve, NHMUK PI BB35212. Scale bar = 10 mm.

3, 4   Interior mould and latex cast of ventral valve, NHMUK PI BB35213. Scale bar = 10 mm.

5   Exterior mould of dorsal valve, NHMUK PI BB35210. Scale bar = 10 mm.

6, 8   Exterior mould and latex cast of ventral valve, NHMUK PI BB35211. Scale bar = 5 mm.

7   Interior mould of ventral valve, NHMUK PI BB35214. Scale bar = 5 mm.

9–13:   *Isophragma parallelum* **n. sp.**

9, 10   Interior mould and latex cast of ventral valve, NHMUK PI BB35252. Scale bar = 2 mm.

11, 12   Holotype. Interior mould and latex cast of dorsal valve, NHMUK PI BB35251. Scale bar = 2 mm.

13   Interior mould of ventral valve, NHMUK PI BB35253. Scale bar = 2 mm.

14, 15:   *Colaptomena pseudopecten?* **(M'Coy, 1846)**

14, 15   Interior mould and latex cast of dorsal valve, NHMUK PI BB35215. Scale bar = 10 mm.

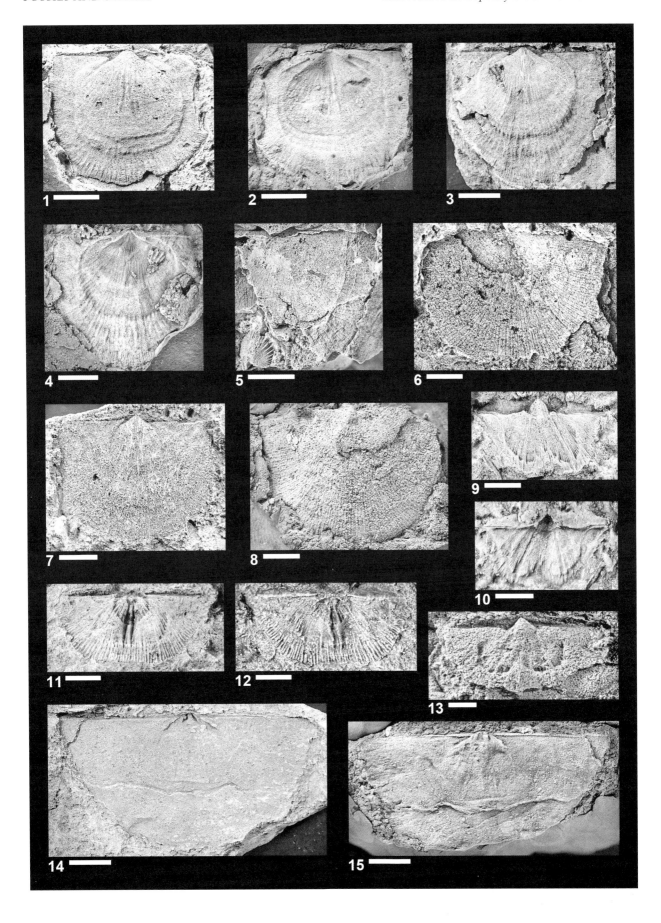

Plate 3

1–4:   *Isophragma parallelum* n. sp.

1, 3   Exterior mould and latex cast of dorsal valve, NHMUK PI BB35249. Scale bar = 2 mm.

2, 4   Exterior mould and latex cast of ventral valve, NHMUK PI BB35250. Scale bar = 2 mm.

5:   *Glyptambonites* sp. Latex cast of ventral valve; scale bar = 5 mm. Specimen lost.

6–14:   *Leptellina llandeiloensis* (Davidson, 1883)

6, 9   Interior mould and latex cast of dorsal valve, NHMUK PI BB35255. Scale bar = 3 mm.

7   Interior mould of dorsal valve, NHMUK PI BB35257. Scale bar = 3 mm.

8   Interior mould of dorsal valve, NHMUK PI BB35256. Scale bar = 3 mm.

10   Interior mould of ventral valve, NHMUK PI BB35259. Scale bar = 3 mm.

11   Interior mould of ventral valve, NHMUK PI BB35258: Scale bar = 3 mm.

12, 13   Exterior mould and latex cast of dorsal valve, NHMUK PI BB35260. Scale bar = 3 mm.

14   Dorsal view of conjoined valves, NHMUK PI BB35254. Scale bar = 3 mm.

15–20: *Grorudia grorudi* Spjeldnæs, 1957

15, 18   Interior mould and latex cast of dorsal valve, NHMUK PI BB35229. Scale bar = 2 mm.

16, 19   Interior mould and latex cast of dorsal bema, NHMUK PI BB35230. Scale bar = 1 mm.

17, 20   Interior mould and latex cast of dorsal valve, NHMUK PI BB35228. Scale bar = 1 mm.

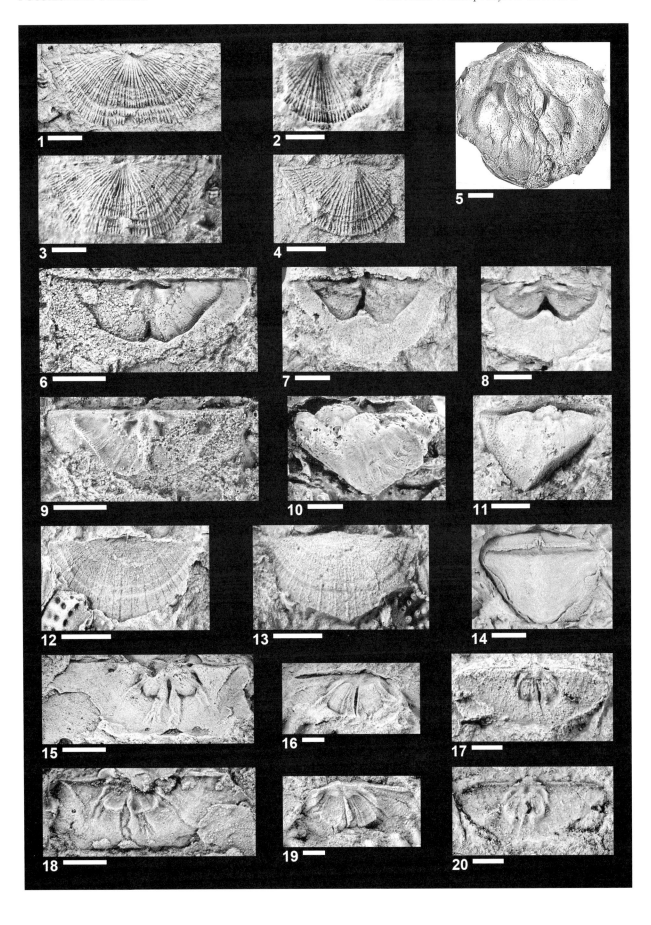

Plate 4

1, 2, 4: *Grorudia grorudi* **Spjeldnæs, 1957**
1, 4   Interior mould and latex cast of ventral valve, NHMUK PI BB35231. Scale bar = 2 mm.
2   Interior mould of ventral valve, NHMUK PI BB35227. Scale bar = 1 mm.

3, 5—9: *Leptestiina derfelensis* **(Jones, 1928)**
3, 5   Interior mould and latex cast of dorsal valve, NHMUK PI BB35261. Scale bar = 1 mm.
6, 9   Interior mould and latex cast of dorsal valve. Specimen lost and valve dimensions unknown.
7, 8   Interior mould and latex cast of ventral valve, NHMUK PI BB35262. Scale bar = 1 mm.

10—17: *Sowerbyella* (*Sowerbyella*) *antiqua* **Jones, 1928**
10, 13   Interior mould and latex cast of dorsal valve, NHMUK PI BB35290. Scale bar = 1 mm.
11, 14   Interior mould and latex cast of dorsal valve, NHMUK PI BB35291. Scale bar = 1 mm.
12, 15   Interior mould and latex cast of dorsal valve, NHMUK PI BB35292. Scale bar = 1 mm.
16   Interior mould of ventral valve, NHMUK PI BB35293. Scale bar = 1 mm.
17   Ventral exterior, NHMUK PI BB35294. Scale bar = 1 mm.

18—21: *Atelelasma longisulcum* **n. sp.**
19, 20   Holotype. Interior mould and latex cast of dorsal valve, NHMUK PI BB35204. Scale bar = 5 mm.
18, 21   Exterior mould and latex cast of dorsal valve, NHMUK PI BB35206. Scale bar = 5 mm.

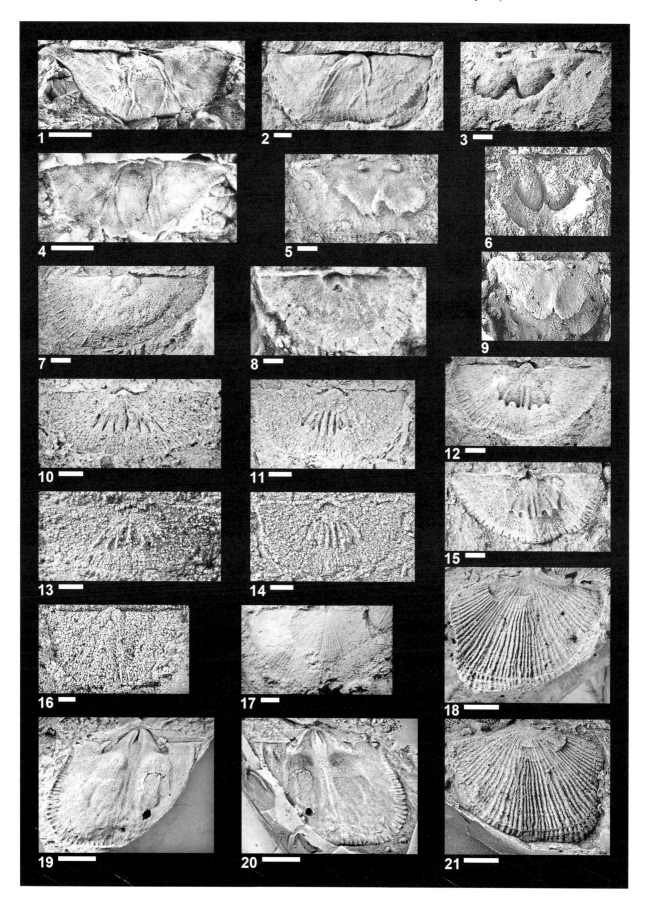

Plate 5

1, 2, 5, 6:	***Atelelasma longisulcum*** **n. sp.**
1, 5	Interior mould and latex cast of ventral valve, NHMUK PI BB35205. Scale bar = 5 mm.
2, 6	Interior mould and latex cast of ventral valve, NHMUK PI BB35207. Scale bar = 5 mm.

3, 4, 7, 8, 11, 12:	***Sulevorthis*** **aff.** ***S. blountensis*** **(Cooper, 1956)**
3	Interior mould of dorsal valve. Scale bar = 2 mm. Specimen lost.
4	Interior mould of dorsal valve. Scale bar = 2 mm. Specimen lost.
7	Interior mould of ventral valve. Scale bar = 2 mm. Specimen lost.
8	Interior mould of ventral valve. Scale bar = 1 mm. Specimen lost.
11	Dorsal exterior. Scale bar = 2 mm. Specimen lost.
12	Dorsal exterior. Scale bar = 1 mm. Specimen lost.

9, 10, 13, 15—20:	***Glyptorthis crispa*** **(M'Coy, 1846)**
9, 10	Interior mould and latex cast of ventral valve, NHMUK PI BB35224. Scale bar = 4 mm.
13	Interior mould of dorsal valve, NHMUK PI BB35226. Scale bar = 3 mm.
15, 18	Interior mould and silicone cast of dorsal valve, NHMUK PI BB35225. Scale bar = 5 mm.
16, 19	Exterior mould and latex cast of dorsal valve, NHMUK PI BB35222. Scale bar = 3 mm.
17, 20	Exterior mould and latex cast of ventral valve, NHMUK PI BB35223. Scale bar = 4 mm.

14:	***Hesperorthis leinsterensis*** **n. sp.** Holotype. Interior mould of dorsal valve, NHMUK PI BB35232. Scale bar = 3 mm.

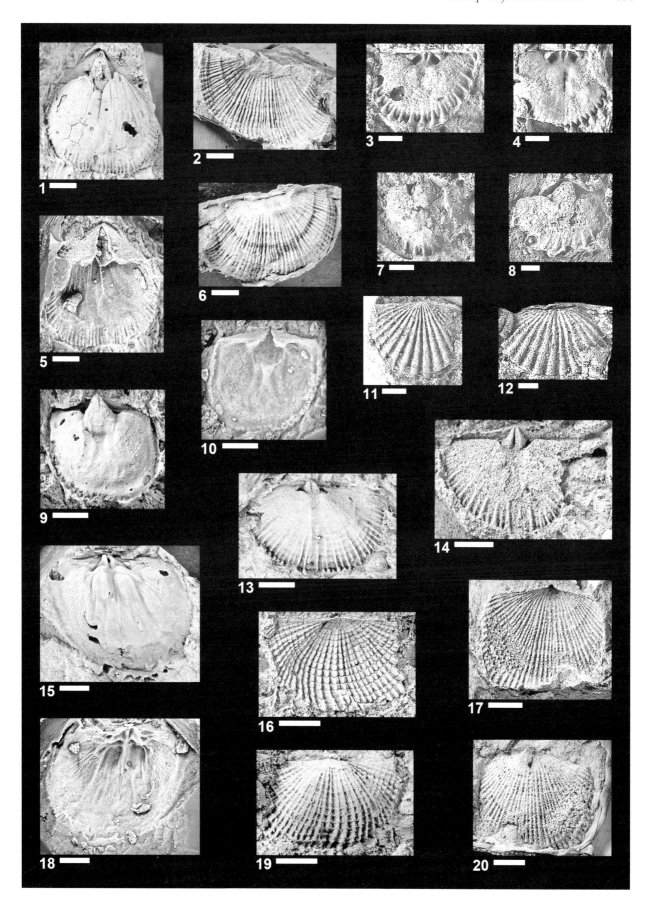

Plate 6

1–13:	***Hesperorthis leinsterensis* n. sp.**
1	Silicone cast of dorsal interior, NHMUK PI BB35232. Scale bar = 3 mm.
2, 3	Interior mould and silicone cast of dorsal valve, NHMUK PI BB35234. Scale bar = 3 mm.
4, 5	Interior mould and silicone cast of ventral valve, NHMUK PI BB35237. Scale bar = 3 mm.
6, 7	Interior mould and silicone cast of ventral valve; NHMUK PI BB35236. Scale bar = 2 mm.
8	Interior mould of ventral valve, NHMUK PI BB35238. Scale bar = 1 mm.
9, 10	Exterior mould and latex cast of dorsal valve, NHMUK PI BB35233. Scale bar = 3 mm.
11	Exterior mould of dorsal valve, NHMUK PI BB35239. Scale bar = 3 mm.
12, 13	Exterior mould and silicone cast of ventral valve, NHMUK PI BB35235. Scale bar = 2 mm.

14–17, 19, 20:	***Valcourea confinis* (Salter, 1849)**
14, 15	Interior mould and silicone cast of dorsal valve, NHMUK PI BB35297. Scale bar = 5 mm.
16, 19	Exterior mould and silicone cast of ventral valve, NHMUK PI BB35295. Scale bar = 5 mm.
17	Interior mould of ventral valve, NHMUK PI BB35296. Scale bar = 5 mm.
20	Latex cast of dorsal exterior. Scale bar = 5 mm. Specimen lost.

21:	***Productorthis* sp.** Interior mould of dorsal valve, NHMUK PI BB35283. Scale bar = 5 mm.

18:	***Platystrophia tramorensis* n. sp.** Anterior view of conjoined valves. Scale bar = 2 mm. Specimen lost.

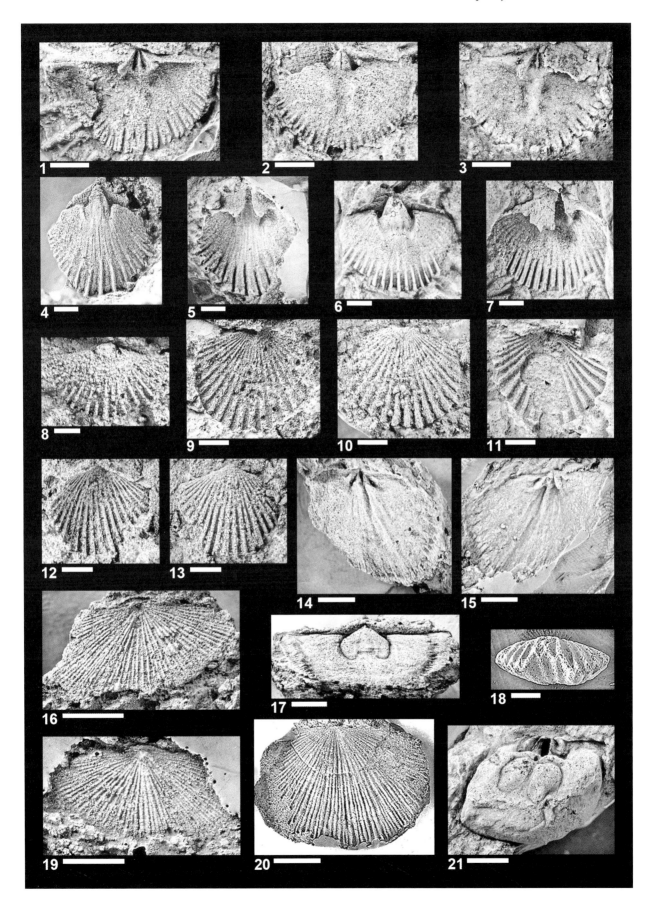

Plate 7

1–9, 11: *Platystrophia tramorensis* **n. sp.**

1, 2   Holotype. Interior mould and silicone cast (anterior view) of dorsal valve; NHMUK PI BB35271. Scale bar = 3 mm.

3   Interior mould of dorsal valve; NHMUK PI BB35276. Scale bar = 3 mm.

4   Interior mould of ventral valve; NHMUK PI BB35274. Scale bar = 3 mm.

5   Interior mould of dorsal valve; NHMUK PI BB35273. Scale bar = 3 mm.

6, 7   Interior mould and silicone cast of ventral valve; NHMUK PI BB35272. Scale bar = 3 mm.

8, 9   Exterior mould and latex cast of dorsal valve; NHMUK PI BB35275. Scale bar = 3 mm.

11   Latex cast of conjoined valves, ventral view; NHMUK PI BB35275. Scale bar = 2 mm.

10, 12–18: *Platystrophia* **aff.** *P. sublimis* Öpik, 1930

10, 13   Interior mould and latex cast of dorsal valve. Scale bar = 3 mm. Specimen lost.

12   Interior mould of ventral valve. Scale bar = 3 mm. Specimen lost.

14   Latex cast of dorsal exterior. Scale bar = 3 mm. Specimen lost.

15   Exterior mould of ventral valve; NHMUK PI BB35269. Scale bar = 3 mm.

16   Latex cast of ventral exterior; NHMUK PI BB35270. Scale bar = 3 mm.

17, 18   Dorsal and anterior view, respectively, of dorsal interior mould; NHMUK PI BB35268. Scale bar = 3 mm.

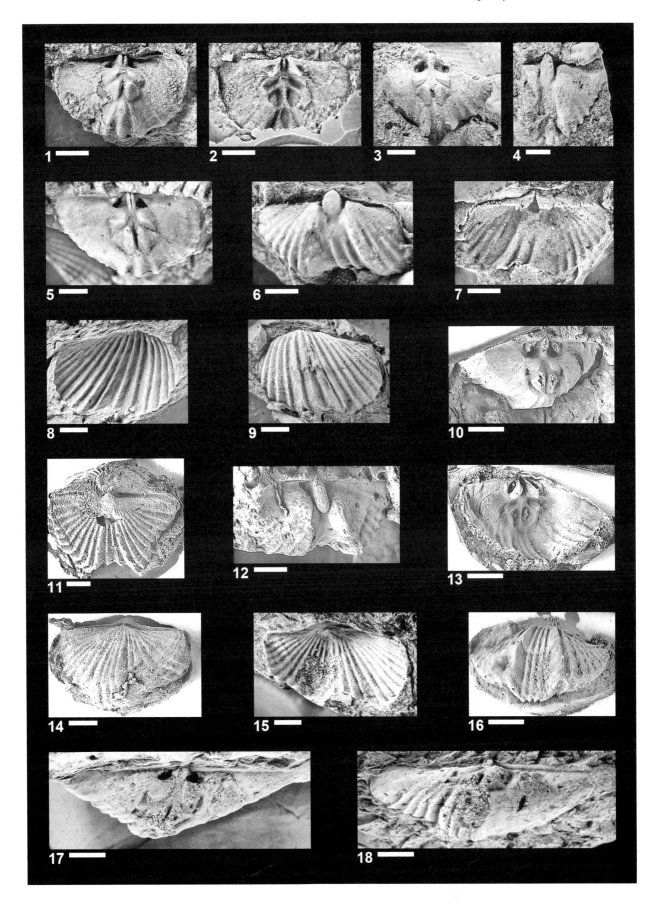

Plate 8

1–11, 13–15: *Howellites hibernicus* **n. sp.**

1	Holotype. Interior mould of dorsal valve; NHMUK PI BB35247. Scale bar = 2 mm.
2, 3	Interior mould and silicone cast of dorsal valve; NHMUK PI BB35242. Scale bar = 2 mm.
4	Interior mould of dorsal valve; NHMUK PI BB35243. Scale bar = 2 mm.
5, 6	Interior mould and latex cast of ventral valve; NHMUK PI BB35245. Scale bar = 2 mm.
7	Interior mould of ventral valve; NHMUK PI BB35240. Scale bar = 2 mm.
8	Interior mould of ventral valve; NHMUK PI BB35241. Scale bar = 2 mm.
9, 10	Exterior mould and latex cast of dorsal valve; NHMUK PI BB35248. Scale bar = 2 mm.
11, 15	Exterior mould and latex cast of ventral valve; NHMUK PI BB35246. Scale bar = 2 mm.
13, 14	Exterior mould and latex cast of dorsal valve; NHMUK PI BB35244. Scale bar = 2 mm.

12:   *Salopia gracilis* **Williams** *in* **Whittington & Williams, 1955** Interior mould of ventral valve; NHMUK PI BB35289. Scale bar = 2 mm.

16–23: *Paurorthis* **aff.** *P. parva* **(Pander, 1830)**

16	Interior mould of dorsal valve; NHMUK PI BB35265. Scale bar = 3 mm.
17, 21	Interior mould and silicone cast of dorsal valve; NHMUK PI BB35266. Scale bar = 2 mm.
18, 22	Interior mould and latex cast of ventral valve; NHMUK PI BB35264. Scale bar = 3 mm.
19, 23	Exterior mould and silicone cast of ventral valve; NHMUK PI BB35263. Scale bar = 3 mm.
20	Interior mould of dorsal valve; NHMUK PI BB35267. Scale bar = 3 mm.

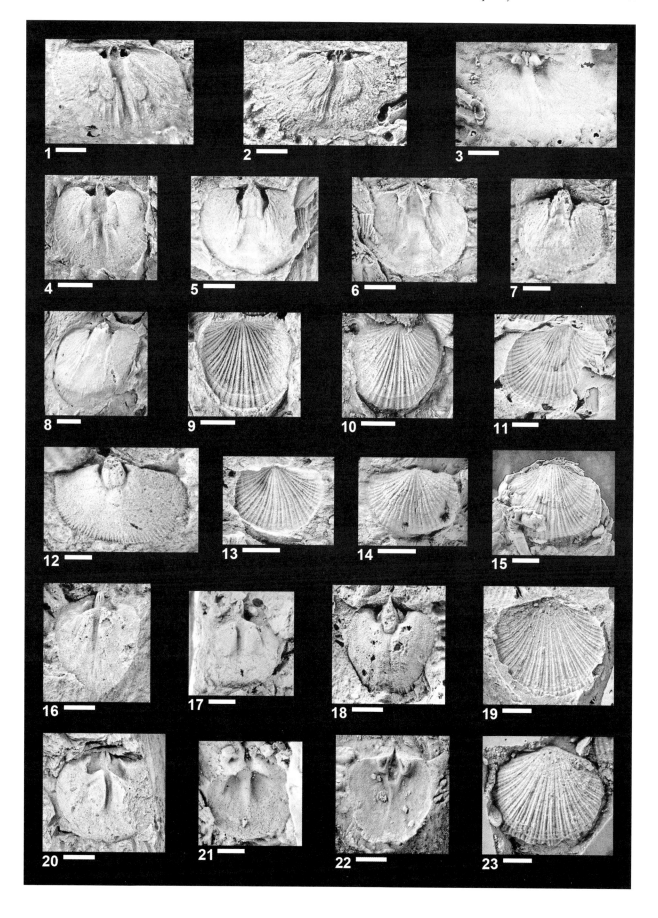

Plate 9

1–9:   *Salopia gracilis* Williams, 1955
1, 5   Interior mould and silicone cast of dorsal valve; NHMUK PI BB35284. Scale bar = 2 mm.
2, 6   Interior mould and latex cast of ventral valve; NHMUK PI BB35286. Scale bar = 2 mm.
3, 7   Exterior mould and silicone cast of dorsal valve; NHMUK PI BB35287. Scale bar = 2 mm.
4, 8   Exterior mould and silicone cast of ventral valve; NHMUK PI BB35288. Scale bar = 2 mm.
9   Interior mould of dorsal valve; NHMUK PI BB35285. Scale bar = 2 mm.

10–17:   *Hibernobonites* n. gen. *filosus* (M'Coy, 1846)
10   Interior mould of dorsal valve; NHMUK PI BB35282. Scale bar = 10 mm.
11   Interior mould of ventral valve; NHMUK PI BB35280. Scale bar = 10 mm.
12, 15   Interior mould and latex cast of dorsal valve; NHMUK PI BB35278. Scale bar = 10 mm.
13, 16   Interior mould and latex cast of ventral valve; NHMUK PI BB35279. Scale bar = 10 mm.
14   Anterior end of ventral sulcus; NHMUK PI BB35277. Scale bar = 10 mm.
17   Latex cast of dorsal exterior mould; NHMUK PI BB35281. Scale bar = 10 mm.

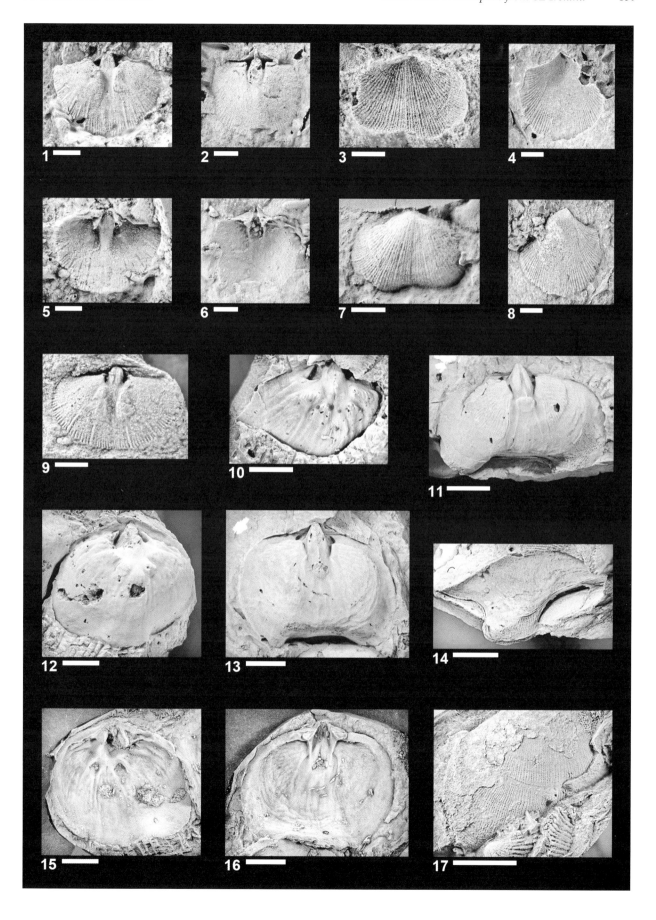

Plate 10

1−4:   *Colaptomena auduni* n. sp.
1   Latex cast of dorsal interior; NHMUK PI BB35208. Scale bar 2 mm.
4   Holotype. Latex cast of ventral interior; NHMUK PI BB35212. Scale bar = 2 mm.

2:   *Dactylogonia costellata* n. sp. Holotype. Latex cast of ventral interior; NHMUK PI BB35216. Scale bar = 2 mm.

3:   *Atelelasma longisulcum* n. sp. Latex cast of ventral interior; NHMUK PI BB35205. Scale bar = 2 mm.

5, 6:   *Isophragma parallelum* n. sp.
5   Holotype. Latex cast of dorsal interior; NHMUK PI BB35251. Scale bar = 2 mm.
6   Latex cast of ventral interior; NHMUK PI BB35252. Scale bar = 2 mm.

7, 9:   *Hesperorthis leinsterensis* n. sp.
7   Silicone cast of dorsal interior; NHMUK PI BB35232. Scale bar = 2 mm.
9   Silicone cast of ventral interior; NHMUK PI BB35237. Scale bar = 2 mm.

8:   *Platystrophia tramorensis* n. sp.
8   Holotype. Silicone cast of dorsal interior; NHMUK PI BB35271. Scale bar = 2 mm.

10−12: *Howellites hibernicus* n. sp.
10   Latex cast of ventral interior; NHMUK PI BB35245. Scale bar = 2 mm.
11   Silicone cast of dorsal interior; NHMUK PI BB35242. Scale bar = 2 mm.
12   Latex cast of dorsal exterior; NHMUK PI BB35248. Scale bar = 2 mm.

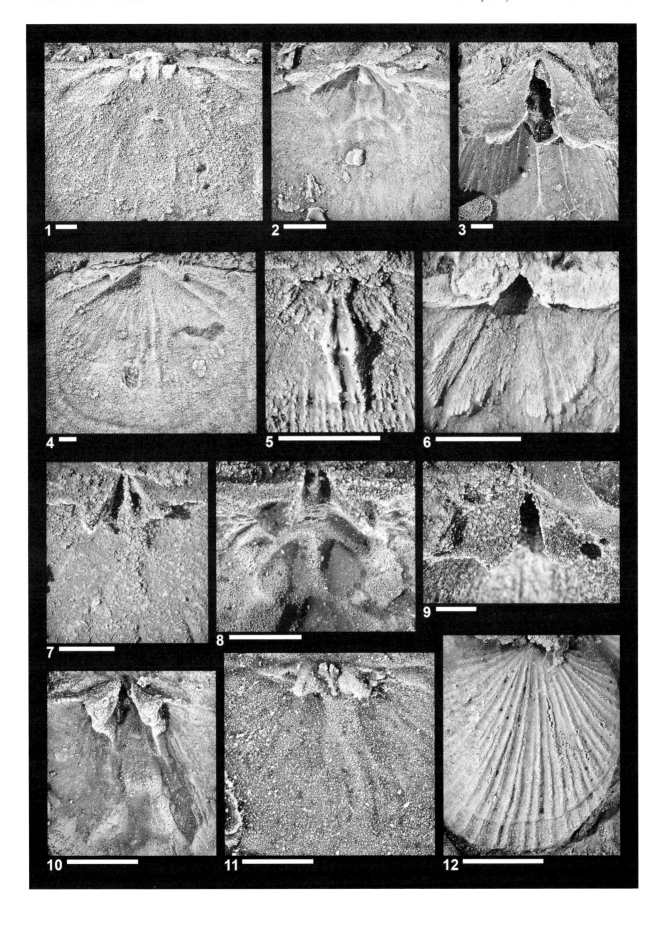

Appendix A

Time-slice (Ma)	Locality	Palaeogeogr.	Country	Area	Stratigraphy	Depth range	British graptol. zones (Cooper & Sadler 2012)	International chronostrat.	Lat.	Long.	Plate ID	References
463	Anglesey	Monian Comp. Terrane	Wales	Anglesey	Nantannog Fm, Bod Deiniol Fm	DR 2–3	Cucullus-murchisoni	Basal–late Darriwilian	53.27	−4.32	376	Neuman & Harper (1992, database)
463	Bellewstown	Bellewstown Terrane	Ireland E	Bellewstown	Hilltown Fm	DR 4	Artus	Early mid Darriwilian	53.64	−6.35	376	Neuman & Harper (1992, database)
463	Bishop's Falls	Central Newfoundland Terrane	Canada	Bishop's Falls, Nfl	Unnamed unit (2a in Neuman 1984)	DR 3–4	Artus? – lowermost gracilis	Early mid Darriwilian – earliest Sandbian	49.05	−55.10	150	Neuman (1984)
463	Bohemia	Perunica	Czech Republic	Bohemia	Dobrotiva Fm, Sarka Fm (shale Mb)	DR 2–3	Upper teretiusculus–lowermost gracilis	Latest Darriwilian – earliest Sandbian	49.92	14.07	374	Havlíček & Vaněk (1966) and Havlíček (1967, 1977)
463	Canada E	Laurentia	Canada E	Québeq S	Laval FM (middle unit)	DR 3	Mid? artus–murchisoni	Mid–late Darriwilian	52.15	−78.89	101	Hofmann (1963)
463	Devon	Armorica	England	Devon	Budleigh Salterton Pebble beds	DR 2–3	Teretiusculus	Latest Darriwilian	50.63	−3.32	305	Cocks & Lockley (1981)
463	Estonia N	Baltica	Estonia N	North Estonia	Kandle Fm (prev. Aseri Fm), Väo Fm, Kõrgekallas Fm	DR 3	Lower murchisoni-teretiusculus	Late mid–latest Darriwilian	59.21	25.68	302	Hints & Rõõmusoks (1997) and Rõõmusoks (2004)
463	Girvan	Midland Valley Terrane	Scotland SW	Girvan	Auchensoul Lst. Mb, Confinis Fm, Stinchar Lst. Fm	DR 3–4	Teretiusculus	Latest Darriwilian	55.21	−4.81	313	Williams (1962)
463	Latvia	Baltica	Latvia W	West Latvia		Shelf	Teretiusculus–gracilis	Latest Darriwilian – early Sandbian	56.88	21.44	302	Hansen (2008)
463	Mayo S	South Mayo Terrane	Ireland W	Mweelrea Mt, South Mayo	Mweelrea Fm	DR 2–3	Murchisoni-teretiusculus	Mid–latest Darriwilian	53.63	−9.83	377	Neuman & Harper (1992, database)
463	Normandy	Armorica	France	Normandy	Gres de May, Gres de petit May	DR 2–3	Teretiusculus	Latest Darriwilian	48.25	−1.66	305	Cocks & Lockley (1981)

(continued)

Appendix A (continued)

Time-slice (Ma)	Locality	Palaeogeogr.	Country	Area	Stratigraphy	Depth range	British graptol. zones (Cooper & Sadler 2012)	International chronostrat.	Lat.	Long.	Plate ID	References
463	Oslo	Baltica	Norway S	Oslo region	Elnes Fm (all mbs)	DR 3–4	Middle artus–teretiusculus	Mid–latest Darriwilian	59.78	10.50	302	Spjeldnæs (1957), Cocks & Rong (1989) and Candela & Hansen (2010)
463	Precordillera	Precordillera	Argentina	San Juan and Jujuy provinces	Las Chacritas Fm, San Juan Fm (uppermost), Sepulturas Fm	DR 2–3 (shallow)	Upper cucullus–teretiusculus	Early–latest Darriwilian	−30.98	−68.89	No plate ID	Benedetto (2003b)
463	Shropshire	Avalonia	England	Shropshire	Weston Flags Fm, Betton Shale Fm, Meadowtown Fm	DR 3–4	Middle murchisoni–teretiusculus	Late Darriwilian	52.59	−2.70	315	Williams (1974)
463	Spain	Iberia	Spain	Zaragoza district	Castillejo Fm (Alpartir Mb), El Castro Fm (lower)	DR 2–3	Lowermost artus–gracilis	Early Darriwilian–early Sandbian	42.40	−3.52	304	Villas (1985)
463	Summerford	Central Newfoundland Terrane	Canada	New World Island, Nfl	Summerford Grp, unnamed Ordovician tuffs	DR 3	Middle murchisoni–teretiusculus	Late Darriwilian	49.05	−55.10	150	Zhan & Jin (2005)
463	Sweden C	Baltica	Sweden C	Siljan district	Seby Fm, Segerstad Fm, Folkeslunda Fm, Furudal Fm	DR 3–4 (shallow)	Lowermost murchisoni–lower gracilis	Late mid Darriwilian–earliest Sandbian	58.36	14.47	302	Jaanusson (1963)
463	Table Head	West Newfoundland Terrane	Canada	Port Saunders, Nfl	Table Point Fm (Table Head Grp.)	DR 3–4 (deep)	Uppermost cucullus–teretiusculus	Early–latest Darriwilian	50.65	−57.29	151	Neuman & Harper (1992, database)
463	Tetagouche	Miramichi Terrane	Canada	Napadogan area, New Brunswick	Tetagouche Grp.	DR 3	Artus-?	Early mid Darriwilian-?	46.43	−66.84	150	Neuman (1984) and Neuman & Harper (1992, database)
463	Tramore	Leinster Terrane	Ireland SE	Tramore Area	Tramore Limestone Fm (Units 1–2)	DR 3	Upper murchisoni–teretiusculus	Late Darriwilian	52.16	−7.13	376	This study

(continued)

Appendix A (continued)

Time-slice (Ma)	Locality	Palaeogeogr.	Country	Area	Stratigraphy	Depth range	British graptol. zones (Cooper & Sadler 2012)	International chronostrat.	Lat.	Long.	Plate ID	References
463	USA E	Laurentia	USA	Tennessee	Lenoir Fm (prev. Arline Fm), Little Oak Fm	DR 3	Lower murchisoni–upper teretiusculus	Late mid–late Darriwilian	36.76	−85.19	101	Cooper (1956) and Zhan & Jin (2005)
463	USA E – deep	Laurentia	USA	Tennessee	Whitesburg Fm, Blockhouse Shale (prev. Athens Fm)	DR 5–6	Middle murchisoni–upper teretiusculus-?	Late Darriwilian	36.76	−85.19	101	Cooper (1956) and Cocks & Rong (1989)
463	USA NE margin	Laurentia	USA	Virginia	Elway Fm, Lenoir Fm (prev. Arline Fm), Little Oak Fm, Red Knobs (Holston) Fm	DR 3 (shallow)	Lower murchisoni–lowermost gracilis	Late mid Darriwilian – earliest Sandbian	41.25	−75.59	101	Cooper (1956), Cocks & Rong (1989) and Zhan & Jin (2005)
463	USA NE margin – deep	Laurentia	USA	Virginia	Whitesburg Fm, Blockhouse Shale (prev. Athens Fm)	DR 5–6	Middle murchisoni–upper teretiusculus-?	Late Darriwilian	41.25	−75.59	101	Cooper (1956)
463	USA SE margin	Laurentia	USA	Alabama	Lenoir Fm (prev. Arline Fm)	DR 3 (shallow)	Lower murchisoni–upper teretiusculus	Late mid–latest Darriwilian	32.32	−86.90	101	Cooper (1956), Cocks & Rong (1989) and Zhan & Jin (2005)
463	Wales S	Avalonia	Wales S	Builth, Llandeilo, Monmouthshire	Ffairfach Group, Lower Llandeilo Flags, Llanfawr mudstones (lower)	DR 2–4	Murchisoni–teretiusculus	Mid–latest Darriwilian	51.94	−3.38	315	Lockley & Williams (1981), Williams et al. (1981) and Cocks (2008)
★	Anglesey Treiorwerth	Monian Comp. Terrane	Wales	Anglesey	Treiorwerth Fm	DR 2–3	Gibberulus–cucullus	Dapingian–early Darriwilian	53.27	−4.32	376	Neuman (1984)
★	Famatina	Famatina Terrane	Argentina	La Rioja Province	Suri Fm (upper Mb)	DR 3–4	Middle varicosus–middle gibberulus	Mid Floian–mid Dapingian	−29.59	−66.83	No plate ID	Benedetto (2003b)
★	Indian Bay	Central Newfoundland Terrane	Canada	Indian Bay big pond, Nfl	Indian Bay Fm	DR 3–4	Cucullus	Early Darriwilian	49.26	−53.75	150	Neuman (1984) and Wonderley & Neuman (1984)

(continued)

Appendix A (continued)

Time-slice (Ma)	Locality	Palaeogeogr.	Country	Area	Stratigraphy	Depth range	British graptol. zones (Cooper & Sadler 2012)	International chronostrat.	Lat.	Long.	Plate ID	References
*	Maine	Miramichi Terrane	USA	Weeksboro and Lunksoos Lake, Maine	Shin Brook Fm	DR 2–3	Cucullus–artus	Basal–late mid Darriwilian	46.17	−68.40	150	Neuman (1964, 1984)
*	Rosslare	Monian Comp. Terrane	Ireland SE	Tagoat village	Tagoat Group	DR 3?	Varicosus–cucullus	Mid Floian–early Darriwilian	52.24	−6.39	376	Neuman & Harper (1992, database)
457	Algeria	Gondwana	Algeria	Algerian Sahara	Azzel Argillite Mb	DR 3–4	Gracilis	Basal–mid Sandbian	33.24	6.43	714	Botquelen & Mélou (2007)
457	Anglesey	Monian Comp. Terrane	Wales N	Anglesey	Garn Fm, Llanbabo Fm, Tandinas Shale	DR 2–4	Gracilis–lower foliaceous	Basal–late Sandbian	53.27	−4.32	376	Bates (1968)
457	Bohemia	Perunica	Czech Republic	Bohemia	Letna Fm, Liben Fm, Vinice Fm (basal ironstone)	DR 2–3	Middle gracilis–lower foliaceous	Early–mid Sandbian	49.92	14.07	374	Havlíček & Vaněk (1966), Havlíček (1967, 1977), Cocks & Rong (1989) and Villas (1992)
457	Brittany	Amorica	France	Amorican Massif, Brittany	Postolonnec Fm (uppermost), Kermeur Fm (basal)	DR 3–4 (shallow)	Upper gracilis–lowermost foliaceous	Mid Sandbian	48.26	−4.60	305	Villas *et al.* (2006) and Botquelen & Mélou (2007)
457	Canada E	Laurentia	Canada E	Quebec	Lowville Fm, Chaumont Fm	DR 3–4 (shallow)	Lower–uppermost gracilis	Earliest–mid Sandbian	52.15	−78.89	101	Cooper (1956)
457	Canada NW	Laurentia	Canada NW	Northwest Territories	Esbataottine Fm	DR 3?	Gracilis	Basal–mid Sandbian	64.83	−124.85	101	Ludvigsen (1975) and Potter & Boucot (1992)
457	Dunabrattin – deep	Leinster Terrane	Ireland SE	Dunabrattin, Bunmahon	Dunabrattin Limestone Fm	DR 4–6	Basal–upper gracilis	Basal–lower mid Sandbian	52.14	−7.28	376	This study

(continued)

Appendix A (continued)

Time-slice (Ma)	Locality	Palaeogeogr.	Country	Area	Stratigraphy	Depth range	British graptol. zones (Cooper & Sadler 2012)	International chronostrat.	Lat.	Long.	Plate ID	References
457	Estonia N	Baltica	Estonia N	North Estonia	Viivikonna Fm (Kukruse oil shales)	DR 3 (shallow)	Gracilis	Basal–mid Sandbian	59.21	25.68	302	Öpik (1930), Cocks & Rong (1989), Hints & Rõõmusoks (1997), Webby *et al.* (1994) and Rõõmusoks (2004)
457	Mayo S	South Mayo Terrane	Ireland W	Mweelrea Mountain, South Mayo	Mweelrea Fm	DR 2–3	Gracilis–lower foliaceous	Basal–late Sandbian	53.63	−9.83	377	Harper *et al.* (2010)
457	Montagne Noire	Amorica	France	Montagne Noire	Kermeur Fm	DR 3–4 (shallow)	Uppermost gracilis–lowermost foliaceous	Mid Sandbian	45.13	3.67	305	Havlíček (1981)
457	Morocco	Gondwana	Morocco	Anti-Atlas Mountains	Izgouiren Fm, Ouinirhen Fm, Tadrist Fm	DR 3?	Lower? gracilis	Earliest? Sandbian	29.97	−8.63	714	Havlíček (1970)
457	Normandy	Amorica	France	Normandy	Chlorite and oolite rich ironstone ~100 m above the base of Vieille-Cour Fm	DR 2–3	Middle gracilis	Early Sandbian	48.25	−1.66	305	Villas (1992)
457	Oslo	Baltica	Norway	Oslo region	Vollen Fm, correlative to Vollen–lower Arnestad fms, Fossum Fm, Furuberget Fm	DR 3–4	Gracilis–lowermost clingani	Sandbian	59.78	10.50	302	Spjeldnaes (1957), Hansen (2008) and Hansen *et al.* (2009)
457	Portugal	Iberia	Portugal	Bucaco	Carregueira Fm (top), Louredo Fm (basal oolitic ironstone)	DR 2–3	Upper gracilis–lowermost foliaceous	Mid Sandbian	40.36	−8.35	304	Botquelen & Mélou (2007) and Villas *et al.* (2011)
457	Precordillera – deep	Precordillera	Argentina	San Juan N	Las Aguaditas Fm	DR 4–5	Gracilis	Basal–mid Sandbian	−30.98	−68.89	No plate ID	Benedetto (2003b)

(continued)

Appendix A (continued)

Time-slice (Ma)	Locality	Palaeogeogr.	Country	Area	Stratigraphy	Depth range	British graptol. zones (Cooper & Sadler 2012)	International chronostrat.	Lat.	Long.	Plate ID	References
457	Shropshire	Avalonia	England	Shropshire	Hoar Edge Grit Fm, Rorrington Shale, Spy Wood Sandstone Fm	DR 3–5?	Uppermost teretiusculus–gracilis	Latest Darriwilian–mid Sandbian	52.59	–2.70	315	Williams (1974), Cocks & Rong (1989) and Cocks (2008)
457	Spain	Iberia	Spain	Zaragoza district	Fombuena Fm (Peña del tomo Mb)	DR 2–3	Middle gracilis	Early Sandbian	41.21	–1.18	304	Villas (1992)
457	Sweden	Baltica	Sweden	Scania, Siljan district, Öland	Sularp Shale, Dalby Lst	DR 2–4	Middle gracilis–upper foliaceous	Earl–latest Sandbian	58.36	14.47	302	Jaanusson (1963), Cocks & Rong (1989), Jaanusson & Bassett (1993), Röõmusoks (2004) and Hansen (2008)
457	Tramore	Leinster Terrane	Ireland	Tramore area	Tramore Limestone Fm, Units 3–5	DR 3 (deep)–4	Gracilis	Basal–mid Sandbian	52.16	–7.13	376	This study
457	USA C	Laurentia	USA	Oklahoma, Missouri, Illinois, Wisconsin	Bromide Fm (Mt. Lake Mb), Camp Nelson Fm, Chaumont Fm	DR 3–4	Gracilis	Basal–mid Sandbian	39.39	–92.41	101	Cooper (1956)
457	USA E	Laurentia	USA	Kentucky, Tennessee	Bromide Fm (Mt. Lake Mb), Camp Nelson Fm, Effna Fm, Elway Fm, Lincolnshire Fm, Poteet Fm, Red Knobs (Holston) Fm	DR 2–4	Upper? teretiusculus–lower? foliaceous	Latest? Darriwilian–late? Sandbian	36.76	–85.19	101	Cooper (1956) and Copper (1977)
457	USA NE margin	Laurentia	USA	Virginia, New York	Camp Nelson Fm, Chaumont Fm, Effna Fm, Lincolnshire Fm, Poteet Fm	DR 2–4	Gracilis–lower foliaceous	Basal–late Sandbian	41.25	–75.59	101	Cooper (1956) and Cocks & Rong (1989)

(continued)

Appendix A (continued)

Time-slice (Ma)	Locality	Palaeogeogr.	Country	Area	Stratigraphy	Depth range	British graptol. zones (Cooper & Sadler 2012)	International chronostrat.	Lat.	Long.	Plate ID	References
457	USA NE margin – shallow+deep	Laurentia	USA	Virginia	Effna-Rich Valley Fm	DR 2–6	Lower gracilis–middle foliaceous	Early–late Sandbian	41.25	−75.59	101	Cooper (1956) and Cocks & Rong (1989)
457	USA SE margin	Laurentia	USA	Alabama	Camp Nelson Fm, Effna Fm	DR 2–4	Lower gracilis–lower foliaceous	Early–late Sandbian	32.32	−86.90	101	Cooper (1956)
457	USA W	Laurentia	USA	Nevada	Bromide Fm (Mt. lake Mb)	DR 3–4	Gracilis	Basal–mid Sandbian	38.80	−116.40	101	Cooper (1956)
457	Wales S	Avalonia	Wales S	Builth, Llandeilo	Llanfawr mudstones and Middle Llandeilo Flags	DR 3–4	Gracilis	Basal–mid Sandbian	51.94	−3.38	315	Lockley & Williams (1981), Williams et al. (1981) and Cocks (2008)

*The locality was not included in the multivariate analyses because the material is significantly older than the material that was statistically compared in the time-slices. The fauna of the locality was only used for manual comparison with the younger Celtic faunas; see Table 6 in the monograph. †Plate ID was not assigned to this plate in the BugPlates program.

Appendix B

Abbreviations in the tables

Abbreviation	Type of measurement	Measured character
A_car	Angle	Cardinalia
A_dp	Angle	Dental plates
B_vm1	Branching	Of *vascula media*, 1. branch
B_vm2	Branching	Of *vascula media*, 2. branch
B_vm3	Branching	Of *vascula media*, 3. branch
B_vm4	Branching	Of *vascula media*, 4. branch
Bema	Description	Of bema
Brach	Mode	Brachiophore bases (convergent, divergent, straight)
C_be	Category	Bema (for PCA)
D	Depth	Valve
Dir	Directions (360 dgr)	Of valves on rock surface
Fulch	Presence/absence	Fulcral plates
L	Length	Valve
L_as	Length	Anterior septa pair
L_be	Length	Bema
L_br	Length	Brachiophores
L_ca	Length	Cardinalia
L_cp	Length	Cardinal process
L_dpl	Length	Dental plates
L_drid	Length	Dorsal ridge
L_gen	Length	To geniculation
L_int	Length	Interarea
L_lr	Length	Lateral ridge
L_mr	Length	Median ridge
L_ms	Length	Median septum
L_msc	Length	Muscle scar
L_my	Length	Myophragm
L_ra	Length	Raised area
L_spo	Length	Spondylium
L_ss	Length	Submedian septa
L_vad	Length	Ventral adductor scar
L_vdi	Length	Ventral diductor scar
W	Width	Valve
W_be	Width	Bema
W_br	Lateral spread	Brachiophores
W_ca	Width	Cardinalia
W_cp	Width	Cardinal process
W_fold	Width	Fold
W_hi	Width	Hingeline
W_lr	Width	Lateral ridge
W_msc	Width	Muscle scar
W_msc	Width	Muscle scar
W_ra	Width	Raised area
W_spo	Width	Spondylium
W_sulc	Width	Sulcus
W_v_add	Width	Ventral adductor scar
W_v_did	Width	Ventral diductor scar

Species (in alphabetical order)

Atelelasma longisulcum n. sp.

Valve	L	W	W_hi	D	L_msc	W_msc	L_mr	W_br	L_int	L_spo	W_spo
Dors.	12.5	19.0	–	–	5.8	10.0	–	8.4	–	–	–
Dors.	14.8	20.5	–	2.1	–	–	–	–	–	–	–
Dors.	15.3	21.1	–	1.7	–	–	–	–	–	–	–
Dors.	16.5	18.0	–	–	7.8	10.0	7.8	9.5	–	–	–
Ventr.	3.5	5.1	4.8	–	–	–	–	–	0.9	0.9	1.0
Ventr.	7.5	11.3	10.4	2.3	–	–	–	–	1.9	3.5	2.8
Ventr.	12.3	15.3	13.2	3.5	–	–	–	–	4.1	4.0	4.5
Ventr.	16.0	20.0	19.5	–	–	–	–	–	5.2	5.5	5.7
Ventr.	17.3	19.2	18.5	–	–	–	–	–	–	–	–
Ventr.	17.8	19.2	17.3	5.3	–	–	–	–	5.7	5.5	6.2
Ventr.	18.1	18.0	–	4.2	–	–	–	–	5.3	5.1	6.6
Ventr.	20.3	21.0	20.3	–	–	–	–	–	4.2	6.5	8.4
Ventr.	21.5	21.0	18.5	6.9	–	–	–	–	6.5	6.4	8.5
Ventr.	–	–	–	–	–	–	–	–	3.0	3.7	4.2

Colaptomena auduni n. sp.

Valve	L	W	D	L_msc	W_msc	L_my	L_lr	W_lr	L_cp	W_cp
Dors.	23.5	28.1	–	–	–	–	7.3	12.5	2.0	2.6
Dors.	27.5	35.0	4.7	–	–	8.5	–	–	3.0	3.3
Dors.	27.5	34.0	–	–	–	11.0	–	–	2.8	3.8
Dors.	31.0	37.2	3.7	–	–	11.5	–	–	3.2	4.0
Dors.	32.5	42.0	–	–	–	13.0	–	–	3.2	4.1
Dors.	33.8	58.0	–	–	–	–	7.2	14.5	2.8	3.8
Dors.	34.0	53.0	–	–	–	–	6.0	11.5	2.5	3.4
Dors.	43.0	64.0	–	–	–	–	–	–	–	–
Ventr.	9.8	15.0	2.5	–	–	–	–	–	–	–
Ventr.	11.5	18.5	–	–	–	–	–	–	–	–
Ventr.	15.0	21.5	–	–	–	–	–	–	–	–
Ventr.	17.2	22.0	–	–	–	–	–	–	–	–
Ventr.	18.2	32.0	–	–	–	–	–	–	–	–
Ventr.	18.5	26.5	–	–	–	–	–	–	–	–
Ventr.	26.7	34.5	–	10.8	11.5	–	–	–	–	–
Ventr.	27.0	33.0	–	–	–	–	–	–	–	–
Ventr.	30.0	33.0	–	11.0	15.5	–	–	–	–	–
Ventr.	31.0	42.0	–	15.5	15.2	–	–	–	–	–
Ventr.	31.0	37.5	–	16.0	20.1	–	–	–	–	–
Ventr.	31.5	–	3.8	–	–	–	–	–	–	–
Ventr.	32.5	41.0	–	12.5	16.0	–	–	–	–	–
Ventr.	33.8	–	–	12.8	15.5	–	–	–	–	–
Ventr.	35.1	44.2	–	14.5	16.5	–	–	–	–	–
Ventr.	38.0	–	4.3	–	–	–	–	–	–	–
Ventr.	38.5	51.0	–	17.0	17.5	–	–	–	–	–
Ventr.	39.1	52.0	–	–	–	–	–	–	–	–
Ventr.	53.1	71.0	3.8	–	–	–	–	–	–	–

Colaptomena pseudopecten? (M'Coy, 1846)

Valve	L	W	L_my	L_ms	L_cp	W_cp
Dors.	31.0	62.0	8.7	14.0	2.0	2.8

Dactylogonia costellata n. sp.

Valve	L	W	D	L_msc	W_msc	L_my	L_cp	W_cp	L_gen	A_dp
Dors.	8.0	11.0	–	–	–	–	–	–	9.2	–
Dors.	9.8	13.2	–	–	–	5.7	–	–	9.5	–
Dors.	10.5	15.0	–	–	–	–	1.0	1.3	9.5	–
Dors.	11.0	16.0	–	–	–	7.1	–	–	9.3	–
Dors.	11.0	14.0	–	–	–	–	–	–	9.0	–
Dors.	11.1	14.5	–	–	–	–	–	–	10.2	–
Dors.	11.5	16.0	–	–	–	–	–	–	9.5	–
Dors.	11.7	15.5	–	–	–	–	1.0	1.5	9.0	–
Dors.	12.0	19.0	–	5.0	5.1	8.2	–	–	–	–
Dors.	12.0	14.0	–	–	–	–	–	–	–	–
Dors.	15.0	22.0	–	5.8	5.1	–	–	–	12.0	–
Ventr.	9.0	–	–	–	–	–	–	–	7.0	–
Ventr.	11.5	17.5	4.3	–	–	–	–	–	–	–
Ventr.	11.8	15.2	–	–	–	–	–	–	9.5	–
Ventr.	12.0	16.5	–	5.0	5.1	–	–	–	10.5	–
Ventr.	14.5	17.5	4.8	5.0	4.5	–	–	–	8.5	94.0
Ventr.	14.8	23.0	–	6.6	5.5	–	–	–	11.2	–
Ventr.	15.0	19.0	–	5.8	5.5	–	–	–	10.0	93.0
Ventr.	15.1	19.5	3.2	5.5	4.7	–	–	–	10.2	–
Ventr.	15.3	21.5	5.2	5.3	6.1	–	–	–	9.5	–
Ventr.	16.0	21.0	5.7	5.5	5.7	–	–	–	–	–
Ventr.	17.4	19.0	–	–	–	–	–	–	–	–

Glyptambonites sp. A

Valve	L	W	L_msc	W_msc	L_ra	W_ra	B_vm1	B_vm2	B_vm3	B_vm4
Ventr.	27.2	35.0	13.4	8.4	7.0	5.0	8.0	12.5	15.5	18.5

Glyptorthis crispa (M'Coy, 1846)

Valve	L	W	L_hi	D	L_msc	W_msc	L_br	W_br	L_dpl
Dors.	1.6	2.7	–	–	–	–	–	–	–
Dors.	3.2	4.5	–	–	–	–	–	–	–
Dors.	4.0	6.7	–	–	–	–	–	–	–
Dors.	6.1	13.2	–	–	–	–	–	–	–
Dors.	6.5	11.5	–	–	–	–	–	–	–
Dors.	7.0	10.2	–	–	–	–	–	–	–
Dors.	7.0	12.0	–	–	–	–	–	–	–
Dors.	7.5	11.0	–	–	–	–	1.3	2.6	–
Dors.	7.5	12.4	12.4	1.5	2.6	4.4	1.2	2.7	–
Dors.	8.2	12.2	–	–	–	–	1.1	2.3	–
Dors.	8.5	14.0	–	–	–	–	–	–	–

(continued)

Glyptorthis crispa (M'Coy, 1846). (continued)

Valve	L	W	L_hi	D	L_msc	W_msc	L_br	W_br	L_dpl
Dors.	8.7	15.0	–	–	–	–	–	–	–
Dors.	9.0	12.8	–	–	–	–	–	–	–
Dors.	9.0	15.0	–	–	–	–	–	–	–
Dors.	9.2	13.0	–	–	–	–	1.3	3.6	–
Dors.	9.2	13.4	12.0	–	–	–	–	–	–
Dors.	9.8	13.4	–	–	–	–	–	–	–
Dors.	10.1	12.8	–	–	2.5	4.2	–	–	–
Dors.	10.5	14.5	12.5	–	–	–	1.7	4.0	–
Dors.	11.0	13.5	10.2	–	4.2	5.4	1.5	3.4	–
Dors.	11.5	15.5	–	–	–	–	1.4	4.0	–
Dors.	11.8	16.0	–	–	–	–	1.7	4.1	–
Dors.	13.0	17.5	–	–	4.0	5.6	–	–	–
Dors.	13.8	15.2	–	–	5.2	5.5	–	–	–
Dors.	13.8	17.5	–	–	–	–	1.8	3.8	–
Dors.	14.2	17.8	16.5	–	–	–	–	–	–
Dors.	14.5	19.5	18.4	–	5.2	7.2	–	–	–
Dors.	14.5	22.5	–	–	–	–	–	–	–
Dors.	15.2	19.5	17.0	–	5.6	7.5	1.6	5.0	–
Dors.	15.5	21.5	–	–	–	–	–	–	–
Dors.	15.8	17.0	–	–	6.2	9.0	1.9	5.1	–
Dors.	16.0	21.0	–	–	–	–	2.5	5.2	–
Dors.	17.5	19.4	19.4	3.1	9.0	11.0	–	–	–
Dors.	17.5	21.0	–	–	–	–	2.9	6.7	–
Dors.	17.5	22.0	–	–	6.4	8.5	–	–	–
Dors.	17.5	22.0	–	–	–	–	–	–	–
Dors.	18.0	23.0	–	–	–	–	2.4	5.8	–
Dors.	19.0	21.0	–	–	7.0	8.5	–	–	–
Dors.	19.0	25.5	–	–	6.0	10.0	–	–	–
Dors.	20.5	23.0	–	–	8.1	8.3	–	–	–
Dors.	–	19.0	–	–	–	–	–	–	–
Dors.	–	20.0	20.0	–	–	–	–	–	–
Dors.	–	–	–	–	7.5	11.5	–	–	–
Ventr.	5.3	9.5	–	–	1.2	1.7	–	–	–
Ventr.	6.0	10.0	–	–	2.3	2.5	–	–	–
Ventr.	7.6	12.8	–	–	–	–	–	–	1.6
Ventr.	8.0	12.0	–	1.4	2.7	2.8	–	–	–
Ventr.	8.2	9.5	–	–	2.7	2.5	–	–	–
Ventr.	9.2	15.5	–	–	2.5	3.0	–	–	–
Ventr.	10.2	13.5	–	–	2.6	2.6	–	–	–
Ventr.	11.0	13.8	–	–	–	–	–	–	2.1
Ventr.	11.5	14.0	12.0	–	4.2	4.0	–	–	1.8
Ventr.	12.0	14.0	–	–	4.0	3.2	–	–	–
Ventr.	12.5	18.5	–	–	–	–	–	–	–
Ventr.	13.0	13.3	11.2	–	4.7	4.2	–	–	3.7
Ventr.	13.0	17.5	–	–	5.0	4.5	–	–	–
Ventr.	13.8	15.2	–	–	6.5	5.3	–	–	–
Ventr.	14.0	19.0	–	–	6.6	5.2	–	–	–
Ventr.	14.4	18.7	18.0	–	–	–	–	–	–
Ventr.	14.5	18.0	–	–	–	–	–	–	3.0
Ventr.	16.0	20.0	–	–	6.6	5.2	–	–	3.5
Ventr.	16.5	18.2	–	–	–	–	–	–	3.2
Ventr.	16.5	18.5	–	–	–	–	–	–	3.6
Ventr.	16.5	19.2	–	–	6.8	6.0	–	–	3.1
Ventr.	16.8	19.8	19.0	2.2	6.0	3.8	–	–	3.7
Ventr.	17.0	19.5	–	–	6.5	5.7	–	–	–
Ventr.	17.0	20.0	–	–	6.0	4.5	–	–	–
Ventr.	17.0	21.0	–	–	–	–	–	–	–
Ventr.	17.2	18.5	–	–	6.3	5.3	–	–	–
Ventr.	17.2	19.5	–	–	–	–	–	–	–
Ventr.	17.2	20.5	–	–	–	–	–	–	–
Ventr.	17.5	16.0			6.0	4.7	–	–	–
Ventr.	17.5	20.0	–	–	5.5	4.5	–	–	–
Ventr.	17.5	22.0	–	–	–	–	–	–	–
Ventr.	18.0	19.0	–	–	7.5	5.3	–	–	–
Ventr.	18.0	19.5	–	–	7.1	5.0	–	–	–
Ventr.	18.0	20.3	–	–	6.2	4.7	–	–	–
Ventr.	18.5	22.0	–	–	2.3	2.7	–	–	–

(continued)

Glyptorthis crispa (M'Coy, 1846). (continued)

Valve	L	W	L_hi	D	L_msc	W_msc	L_br	W_br	L_dpl
Ventr.	18.5	22.3	19.4	–	–	–	–	–	–
Ventr.	18.5	22.5	20.0	–	7.0	5.5	–	–	–
Ventr.	20.0	22.5	–	–	–	–	–	–	–
Ventr.	20.5	22.0	–	–	7.5	6.4	–	–	3.8
Ventr.	23.0	25.0	–	–	8.5	6.5	–	–	–
Ventr.	–	15.0	–	–	–	–	–	–	–
Ventr.	–	16.0	–	–	–	–	–	–	–
Ventr.	–	24.0	24.0				–	–	–
Ventr.	–	–	–	–	6.3	5.5	–	–	–
Ventr.	–	–	–	–	–	–	–	–	–

Grorudia grorudi Spjeldnæs, 1957

Valve	L	W	D	L_msc	W_msc	B_vm1
Dors.	2.3	7.5	–	1.3	2.0	–
Dors.	2.3	7.5	–	1.3	2.0	–
Dors.	2.6	8.5	–	1.5	1.8	–
Dors.	2.6	8.4	–	1.6	2.2	–
Dors.	2.6	8.2	–	1.9	3.1	–
Dors.	2.7	10.2	–	1.2	1.7	–
Dors.	3.0	15.0	–	1.5	3.0	–
Dors.	3.2	9.2	–	1.3	1.7	–
Dors.	3.6	10.4	–	1.8	3.0	–
Dors.	3.6	10.2	–	2.1	2.5	–
Dors.	–	–	–	2.0	3.1	–
Dors.	–	–	–	1.7	2.5	–
Dors.	–	–	–	1.9	2.2	–
Ventr.	2.1	6.3	–	–	–	–
Ventr.	2.3	7.5	–	–	–	–
Ventr.	2.3	4.5	–	–	–	–
Ventr.	2.6	8.5	–	–	–	–
Ventr.	2.8	6.5	–	–	–	–
Ventr.	3.0	6.7	–	–	–	–
Ventr.	3.0	6.0	–	–	–	–
Ventr.	3.0	5.5	–	–	–	–
Ventr.	3.2	11.8	–	–	–	–
Ventr.	3.2	10.5	–	–	–	2.8
Ventr.	3.3	14.0	–	–	–	3.2
Ventr.	3.4	12.0	–	–	–	–
Ventr.	3.4	7.5	–	–	–	2.2
Ventr.	3.5	9.0	–	–	–	–
Ventr.	3.5	7.8	–	–	–	–
Ventr.	3.6	10.1	–	–	–	2.1
Ventr.	3.7	14.5	–	–	–	–
Ventr.	3.8	9.2	–	–	–	2.6
Ventr.	4.2	11.5	–	–	–	2.7
Ventr.	4.3	9.6	–	–	–	2.3
Ventr.	4.3	9.5	–	–	–	–
Ventr.	4.4	12.5	–	–	–	–
Ventr.	4.5	13.5	–	–	–	3.0
Ventr.	4.5	12.4	–	–	–	–
Ventr.	4.6	18.8	–	–	–	–
Ventr.	4.6	9.8	–	–	–	–
Ventr.	4.7	12.0	–	–	–	2.4
Ventr.	4.8	11.4	–	–	–	–
Ventr.	4.9	–	–	–	–	2.2
Ventr.	5.2	16.5	–	–	–	–
Ventr.	5.2	15.0	–	–	–	–
Ventr.	6.0	19.8	–	–	–	3.4

Hesperorthis leinsterensis n. sp.

Valve	L	W	D	L_msc	W_msc	L_mr	L_br	W_br	L_int
Dors.	4.0	6.4	–	–	–	–	–	–	–
Dors.	4.4	6.5	–	–	–	–	–	–	–
Dors.	5.0	6.8	–	–	–	–	–	–	–
Dors.	5.1	8.4	–	–	–	3.8	1.2	1.4	–
Dors.	5.6	7.5	–	–	–	4.0	1.1	1.3	–
Dors.	5.8	5.6	–	–	–	4.5	1.2	1.7	–
Dors.	6.4	8.5	–	–	–	–	–	–	–
Dors.	6.6	9.5	–	2.5	3.0	5.0	1.6	1.9	–
Dors.	7.0	9.1	–	–	–	–	–	–	–
Dors.	7.1	8.9	–	–	–	–	–	–	–
Dors.	7.5	9.4	–	–	–	–	1.5	2.3	–
Dors.	7.5	11.5	–	–	–	5.5	1.7	2.5	–
Dors.	8.0	11.6	–	2.5	2.9	–	1.9	2.7	–
Dors.	8.0	10.6	–	3.2	3.5	–	2.0	2.7	–
Dors.	8.1	10.5	–	–	–	–	–	–	–
Dors.	8.5	8.7	–	–	–	–	–	–	–
Dors.	8.9	10.4	–	–	–	–	2.0	2.9	–
Dors.	10.0	13.2	2.0	–	–	–	–	–	–
Dors.	10.4	13.1	–	–	–	–	–	–	–
Ventr.	2.2	3.4	–	0.6	0.8	–	–	–	–
Ventr.	3.0	4.9	–	0.8	1.1	–	–	–	0.4
Ventr.	3.1	5.1	–	0.8	1.0	–	–	–	–
Ventr.	3.5	6.3	–	1.0	1.1	–	–	–	0.5
Ventr.	4.3	6.2	–	–	–	–	–	–	–
Ventr.	4.7	7.0	–	1.4	1.3	–	–	–	0.6
Ventr.	4.8	5.5	1.8	–	–	–	–	–	–
Ventr.	5.0	6.5	–	–	–	–	–	–	–
Ventr.	5.1	7.4	2.2	–	–	–	–	–	–
Ventr.	5.5	7.5	1.7	–	–	–	–	–	–
Ventr.	5.7	7.2	–	2.1	1.8	–	–	–	1.0
Ventr.	5.7	8.6	–	1.7	1.4	–	–	–	0.8
Ventr.	5.9	7.2	1.7	–	–	–	–	–	–
Ventr.	5.9	8.8	–	–	–	–	–	–	1.0
Ventr.	6.0	6.6	–	1.7	1.3	–	–	–	–
Ventr.	7.0	8.6	–	–	–	–	–	–	–
Ventr.	7.1	8.8	2.3	–	–	–	–	–	–
Ventr.	7.2	8.5	–	2.5	2.0	–	–	–	1.7
Ventr.	7.3	8.5	–	2.7	1.9	–	–	–	1.1
Ventr.	7.5	8.8	–	2.4	1.8	–	–	–	1.7
Ventr.	8.0	10.2	3.7	–	–	–	–	–	–
Ventr.	8.2	10.3	–	2.6	1.7	–	–	–	1.5
Ventr.	8.3	10.2	–	–	–	–	–	–	2.0
Ventr.	8.8	9.8	–	3.0	2.7	–	–	–	1.7
Ventr.	8.8	9.1	–	2.2	1.7	–	–	–	–
Ventr.	9.3	11.0	–	–	–	–	–	–	1.7
Ventr.	9.6	11.3	–	3.4	2.5	–	–	–	2.3
Ventr.	10.0	11.6	3.9	–	–	–	–	–	–
Ventr.	10.2	10.8	–	3.7	2.4	–	–	–	2.3
Ventr.	12.7	11.6	–	–	–	–	–	–	3.5
Ventr.	13.0	13.5	–	5.0	3.5	–	–	–	–
Ventr.	14.2	12.4	–	5.8	3.3	–	–	–	3.2
Ventr.	15.0	17.0	–	5.1	4.1	–	–	–	–

Hibernobonites n. gen. *filosus* (M'Coy, 1846) (Table 1)

Valve	L	W	L_msc	W_msc	L_cp	L_br	W_br	L_drid	W_fold
Dors.	36.0	42.5	8.1	17.3	–	7.8	10.1	12.0	17.0
Dors.	32.9	40.3	–	–	–	–	–	11.5	–
Dors.	–	–	7.8	15.5	2.2	5.5	9.0	16.2	–
Dors.	–	–	10.5	17.8	2.8	8.1	10.8		–

Hibernobonites n. gen. *filosus* (M'Coy, 1846) (Table 2)

Valve	L	W	W_hi	L_msc	W_msc	L_vdi	W_vdi	L_vad	W_vad	L_dpl
Ventr.	31.5	40.5	15.0	9.5	5.8	4.3	4.7	11.5	14.2	21.0
Ventr.	26.0	38.5	15.4	9.0	5.2	4.0	4.8	10.0	13.5	20.0
Ventr.	29.5	39.0	–	–	–	–	–	–	–	20.0
Ventr.	27.0	32.3	–	8.3	4.5	–	–	9.5	13.5	16.0
Ventr.	30.0	34.5	–	–	–	–	–	–	–	18.5
Ventr.	32.0	–	–	–	–	4.3	4.0	–	–	–
Ventr.	31.0	–	–	–	–	4.5	3.5	–	–	–

Howellites hibernicus n. sp.

Sample	Valve	L	W	D	L_msc	W_msc	L_dpl	Brach	Fulch
A	Dors.	4.5	6.4	0.5	1.6	1.9	–	–	–
A	Dors.	5.2	7.3	0.3	2.3	3.2	–	Straight	–
A	Dors.	5.3	7.2	0.8	2.3	2.9	–	–	–
A	Dors.	5.3	6.5	0.4	2.0	2.5	–	–	–
A	Dors.	6.0	8.5	0.7	2.6	2.8	–	–	–
A	Dors.	7.0	8.4	0.6	2.3	2.7	–	–	–
A	Dors.	7.2	7.3	0.5	2.9	2.8	–	–	–
A	Dors.	7.3	8.4	0.5	3.1	3.5	–	–	–
A	Dors.	7.3	9.5	0.6	3.1	3.5	–	–	–
A	Dors.	7.3	8.6	0.5	3.4	3.5	–	Divergent	–
A	Dors.	7.6	8.4	0.6	2.3	2.7	–	–	–
A	Dors.	7.7	7.9	0.6	3.3	3.2	–	–	–
A	Dors.	8.1	7.7	0.6	2.8	2.8	–	–	–
A	Ventr.	6.3	5.5	1.3	2.2	1.2	–	–	–
A	Ventr.	6.5	7.0	0.9	2.4	1.7	–	–	–
A	Ventr.	7.1	8.2	1.3	2.5	2.5	–	–	–
A	Ventr.	7.2	7.5	0.7	2.3	2.1	–	–	–
A	Ventr.	7.2	9.2	0.7	2.3	2.1	–	–	–
A	Ventr.	7.2	8.6	0.6	2.4	2.3	–	–	–
A	Ventr.	7.8	7.6	0.9	1.9	1.4	–	–	–
A	Ventr.	7.9	8.3	1.2	3.0	2.0	–	–	–
A	Ventr.	7.9	9.4	0.8	2.9	2.3	–	–	–
A	Ventr.	8.1	7.2	0.8	2.0	1.5	–	–	–
A	Ventr.	8.7	10.7	0.5	3.1	2.3	–	–	–
A	Ventr.	9.3	8.5	1.5	2.7	1.5	–	–	–
B	Dors.	3.7	4.3	0.8	1.6	1.6	–	Straight	–
B	Dors.	3.7	6.0	0.8	1.7	2.6	–	–	–
B	Dors.	3.9	5.1	0.6	1.8	1.7	–	–	–
B	Dors.	4.0	7.2	0.3	2.1	2.5	–	–	–
B	Dors.	4.0	6.5	0.7	1.8	2.4	–	Divergent	–
B	Dors.	4.1	5.5	0.4	2.5	2.4	–	Divergent	–
B	Dors.	4.2	5.3	0.5	1.9	2.3	–	Straight	–
B	Dors.	4.2	5.6	0.4	1.8	2.2	–	–	–
B	Dors.	4.2	4.1	0.4	1.6	1.9	–	Straight	–
B	Dors.	4.2	5.6	0.4	1.8	1.8	–	–	–
B	Dors.	4.2	6.2	–	1.7	2.5	–	–	–
B	Dors.	4.3	6.4	0.3	1.7	2.6	–	Convergent	–
B	Dors.	4.3	6.6	0.6	2.6	2.5	–	–	Present
B	Dors.	4.3	7.0	0.5	2.1	2.5	–	–	–
B	Dors.	4.3	5.3	0.5	1.9	2.5	–	Divergent	–
B	Dors.	4.5	5.7	0.4	2.1	2.0	–	–	–
B	Dors.	4.5	5.5	0.6	1.9	2.5	–	–	Present
B	Dors.	4.6	5.2	0.6	2.5	2.2	–	–	–
B	Dors.	4.6	5.3	0.5	1.8	2.0	–	Straight	–
B	Dors.	4.6	5.2	0.4	2.3	2.2	–	Divergent	–

(continued)

Howellites hibernicus n. sp. (continued)

Sample	Valve	L	W	D	L_msc	W_msc	L_dpl	Brach	Fulch
B	Dors.	4.6	5.5	0.6	1.9	2.5	–	–	–
B	Dors.	4.7	5.1	0.4	2.1	2.3	–	Divergent	–
B	Dors.	4.7	6.3	0.6	2.1	2.4	–	Straight	–
B	Dors.	4.7	7.7	0.8	2.1	3.2	–	Straight	–
B	Dors.	4.7	5.0	0.7	2.0	2.3	–	Convergent	–
B	Dors.	4.8	5.0	0.5	2.6	2.1	–	Straight	–
B	Dors.	5.1	8.1	0.5	2.2	2.7	–	–	–
B	Dors.	5.2	6.1	0.6	2.0	2.6	–	Convergent	–
B	Dors.	5.2	5.4	0.8	2.1	2.4	–	Straight	–
B	Dors.	5.2	5.6	0.6	2.8	2.6	–	–	–
B	Dors.	5.2	8.2	0.7	2.7	3.0	–	Straight	(Present)
B	Dors.	5.2	5.6	0.5	2.3	1.9	–	–	–
B	Dors.	5.2	7.2	1.0	2.4	2.5	–	Divergent	–
B	Dors.	5.2	8.1	0.8	2.4	3.4	–	Straight	–
B	Dors.	5.3	6.6	0.3	2.3	2.8	–	Straight	–
B	Dors.	5.3	7.8	0.5	2.3	2.5	–	–	–
B	Dors.	5.4	5.4	0.4	2.3	2.5	–	Straight	–
B	Dors.	5.5	8.1	0.7	3.0	4.2	–	–	–
B	Dors.	5.6	6.9	0.5	3.1	3.1	–	Straight	–
B	Dors.	5.7	7.7	0.4	2.5	2.8	–	–	–
B	Dors.	5.8	6.3	0.6	1.9	2.4	–	Divergent	–
B	Dors.	5.9	8.5	0.4	2.6	3.4	–	Straight	Present
B	Dors.	6.0	6.2	0.5	2.7	2.6	–	Divergent	–
B	Dors.	6.0	7.5	0.4	2.5	3.1	–	–	–
B	Dors.	6.1	6.7	0.7	2.8	3.2	–	Convergent	–
B	Dors.	6.1	8.5	0.5	2.7	3.0	–	–	–
B	Dors.	6.2	7.0	0.4	2.6	3.4	–	–	–
B	Dors.	6.2	8.1	0.6	2.9	3.5	–	–	–
B	Dors.	6.9	7.0	0.7	3.2	2.3	–	Divergent	Present
B	Dors.	7.1	8.9	0.5	2.5	3.4	–	Straight	Present
B	Dors.	3.7	4.0	0.3	–	–	–	Straight	–
B	Dors.	4.8	5.7	0.8	–	–	–	Divergent	–
B	Dors.	6.2	6.1	0.5	–	–	–	Divergent	Present
B	Dors.	6.5	7.5	1.0	–	–	–	–	–
B	Ventr.	4.1	5.3	1.1	2.0	1.5	–	–	–
B	Ventr.	4.2	5.6	0.9	1.6	1.5	–	–	–
B	Ventr.	4.3	5.7	0.9	1.3	1.4	–	–	–
B	Ventr.	4.4	5.6	1.2	1.7	1.6	–	–	–
B	Ventr.	4.5	6.7	1.3	1.5	1.7	–	–	–
B	Ventr.	4.5	6.7	1.3	1.5	1.7	1.4	–	–
B	Ventr.	4.6	5.2	1.4	1.5	1.3	–	–	–
B	Ventr.	4.7	5.3	1.3	2.1	1.4	–	–	–
B	Ventr.	4.7	5.2	0.9	1.3	1.5	–	–	–
B	Ventr.	4.7	4.5	1.2	2.0	1.6	1.6	–	–
B	Ventr.	4.7	6.0	1.3	2.0	1.5	1.2	–	–
B	Ventr.	4.7	5.0	1.3	1.4	1.5	–	–	–
B	Ventr.	4.8	7.9	1.2	1.9	1.7	–	–	–
B	Ventr.	5.0	5.3	1.3	2.2	1.8	–	–	–
B	Ventr.	5.0	5.3	1.2	2.2	1.7	–	–	–
B	Ventr.	5.2	6.0	1.8	2.2	1.5	2.2	–	–
B	Ventr.	5.2	7.0	1.1	1.8	1.7	–	–	–
B	Ventr.	5.3	8.2	1.3	1.7	1.7	–	–	–
B	Ventr.	5.4	5.3	0.8	2.1	1.7	–	–	–
B	Ventr.	5.4	5.6	1.4	2.3	1.5	–	–	–
B	Ventr.	5.4	6.8	1.2	2.0	1.8	–	–	–
B	Ventr.	5.4	7.5	1.4	1.4	2.3	0.4	–	–
B	Ventr.	5.4	6.3	1.6	2.0	1.6	–	–	–
B	Ventr.	5.4	6.1	1.3	2.4	1.5	–	–	–
B	Ventr.	5.5	5.3	1.1	1.8	1.4	–	–	–
B	Ventr.	5.6	4.8	1.0	1.7	1.4	–	–	–
B	Ventr.	5.7	7.7	0.4	2.5	2.8	–	–	–
B	Ventr.	5.7	6.7	0.7	2.1	1.9	1.1	–	–
B	Ventr.	5.7	6.7	1.3	1.5	2.0	1.5	–	–
B	Ventr.	5.7	5.3	1.2	2.0	1.7	–	–	–
B	Ventr.	5.8	5.3	1.2	2.0	1.4	–	–	–
B	Ventr.	5.8	7.2	1.1	2.4	1.6	1.3	–	–
B	Ventr.	5.9	5.8	1.2	2.2	1.4	–	–	–

(continued)

Howellites hibernicus n. sp. (continued)

Sample	Valve	L	W	D	L_msc	W_msc	L_dpl	Brach	Fulch
B	Ventr.	5.9	6.7	1.2	1.9	1.8	–	–	–
B	Ventr.	5.9	8.3	0.9	2.5	1.8	–	–	–
B	Ventr.	6.0	6.0	1.2	1.9	1.7	–	–	–
B	Ventr.	6.0	7.2	1.1	2.3	1.6	–	–	–
B	Ventr.	6.0	6.3	1.1	2.1	1.5	–	–	–
B	Ventr.	6.1	6.7	1.2	2.4	1.7	–	–	–
B	Ventr.	6.1	6.3	1.1	2.1	1.5	–	–	–
B	Ventr.	6.2	6.0	1.2	1.6	1.3	–	–	–
B	Ventr.	6.2	7.0	0.9	1.9	2.2	–	–	–
B	Ventr.	6.3	6.8	1.3	2.1	1.7	–	–	–
B	Ventr.	6.5	6.8	1.2	2.3	1.7	–	–	–
B	Ventr.	6.5	8.4	1.3	2.3	2.2	–	–	–
B	Ventr.	6.7	6.5	0.7	2.3	1.8	–	–	–
B	Ventr.	6.8	7.2	1.1	2.4	1.6	1.3	–	–
B	Ventr.	6.9	9.0	1.4	2.7	1.9	–	–	–
B	Ventr.	7.2	7.2	1.5	2.3	1.8	–	–	–
B	Ventr.	7.3	7.2	1.4	2.9	1.5	–	–	–
B	Ventr.	7.7	6.4	1.8	2.8	1.7	–	–	–
B	Ventr.	7.7	8.8	0.7	2.5	2.0	–	–	–
B	Ventr.	7.8	7.1	1.2	2.6	2.1	–	–	–
B	Ventr.	8.2	8.0	1.2	2.9	2.0	–	–	–
B	Ventr.	3.5	5.2	1.3	–	–	–	–	–
B	Ventr.	5.0	7.2	1.5	–	–	–	–	–
B	Ventr.	5.3	8.1	1.9	–	–	–	–	–
B	Ventr.	7.3	8.6	2.2	–	–	–	–	–
Other	Dors.	5.4	6.0	–	–	–	–	–	–
Other	Ventr.	5.8	6.1	–	1.7	1.1	1.5	–	–
Other	Ventr.	6.2	8.1	–	–	–	–	–	–

Isophragma parallelum n. sp.

Valve	L	W	L_msc	W_msc	L_ss	L_be	W_be	L_br	W_br
Dors.	2.9	6.9	1.5	1.9	2.3	–	–	0.3	1.5
Dors.	2.9	5.9	–	–	2.5	–	–	–	–
Dors.	3.0	7.0	1.3	1.7	–	–	–	0.4	1.6
Dors.	3.4	10.9	1.7	2.5	2.9	–	–	0.4	2.4
Dors.	4.1	9.5	1.8	2.1	3.2	–	–	0.4	2.0
Ventr.	2.5	5.2	0.6	0.9	–	–	–	–	–
Ventr.	3.5	6.8	1.0	1.2	–	2.5	3.4	–	–
Ventr.	3.7	8.0	–	–	–	–	–	–	–
Ventr.	3.8	7.3	0.9	1.2	–	3.4	4.0	–	–
Ventr.	5.3	13.5	1.6	2.0	–	–	–	–	–
Ventr.	5.7	12.5	1.4	2.1	–	4.0	6.5	–	–

Leptellina llandeiloensis (Davidson, 1883)

Valve	L	W	D	L_msc	W_msc	L_ca	W_ca	L_be	W_be
Dors.	2.6	5.9	–	–	–	–	–	–	–
Dors.	3.8	9.2	–	–	–	0.5	2.2	2.6	7.1
Dors.	3.8	8.5	–	–	–	0.4	2.2	2.4	6.0
Dors.	4.9	8.2	1.0	–	–	–	–	–	–
Dors.	5.1	12.2	–	–	–	0.8	2.7	3.8	8.3
Dors.	5.3	11.8	–	–	–	0.9	2.9	3.7	8.8
Dors.	5.3	9.4	–	–	–	0.8	2.6	3.5	6.5
Dors.	5.3	9.2	–	–	–	–	–	–	–

(continued)

Leptellina llandeiloensis (Davidson, 1883). (continued)

Valve	L	W	D	L_msc	W_msc	L_ca	W_ca	L_be	W_be
Dors.	5.5	–	–	–	–	1.1	2.9	3.4	9.2
Dors.	5.6	12.4	–	–	–	–	–	–	–
Dors.	5.7	11.2	–	–	–	–	3.6	3.4	7.8
Dors.	5.8	10.4	–	–	–	0.7	3.8	3.9	8.6
Dors.	6.3	16.1	–	–	–	0.8	4.7	4.7	11.7
Dors.	6.3	10.3	2.1	–	–	–	–	–	–
Dors.	6.3	8.6	–	–	–	–	–	–	–
Dors.	6.4	10.6	–	–	–	–	–	–	–
Dors.	6.7	9.5	–	–	–	–	–	–	–
Dors.	7.2	13.2	–	–	–	0.9	3.8	4.6	11.1
Dors.	7.2	11.3	2.2	–	–	–	–	–	–
Dors.	7.3	–	–	–	–	1.1	4.2	5.2	10.5
Dors.	8.2	11.2	2.5	–	–	–	–	–	–
Dors.	9.7	15.5	2.4	–	–	–	–	–	–
Dors.	–	–	–	–	–	1.1	3.9	4.7	9.2
Dors.	–	12.1	–	–	–	0.8	3.6	4.1	10.1
Dors.	–	–	–	–	–	0.7	3.8	4.8	10.9
Dors.	–	–	–	–	–	0.8	3.9	3.3	10.8
Dors.	–	–	–	–	–	0.6	2.9	3.4	8.3
Dors.	–	–	–	–	–	0.7	3.8	3.7	9.2
Dors.	–	–	–	–	–	0.8	3.8	4.2	10.6
Dors.	–	–	–	–	–	0.9	3.9	4.9	9.8
Dors.	–	–	–	–	–	1.0	2.8	4.2	8.5
Dors.	–	13.5	–	–	–	1.1	4.2	3.8	8.9
Dors.	–	–	–	–	–	1.2	4.6	5.2	12.3
Dors.	–	–	–	–	–	0.9	4.4	6.1	13.0
Dors.	–	–	–	–	–	1.0	2.8	3.6	7.5
Ventr.	4.7	10.3	1.4	1.4	2.0	–	–	–	–
Ventr.	5.9	10.8	3.1	2.1	2.7	–	–	–	–
Ventr.	6.3	11.1	3.6	1.7	2.8	–	–	–	–
Ventr.	6.4	11.0	2.7	1.7	2.7	–	–	–	–
Ventr.	6.7	10.0	3.9	–	–	–	–	–	–
Ventr.	6.8	11.2	3.5	1.6	2.6	–	–	–	–
Ventr.	6.8	10.4	4.2	1.5	2.8	–	–	–	–
Ventr.	6.9	8.7	2.9	1.7	2.3	–	–	–	–
Ventr.	7.1	9.4	2.5	1.8	2.7	–	–	–	–
Ventr.	7.1	11.2	3.3	–	–	–	–	–	–
Ventr.	7.5	12.4	–	–	–	–	–	–	–
Ventr.	7.6	10.1	3.5	2.6	2.6	–	–	–	–
Ventr.	8.1	14.3	–	2.2	3.3	–	–	–	–
Ventr.	8.2	11.5	3.9	2.5	2.7	–	–	–	–
Ventr.	8.9	13.6	4.6	–	–	–	–	–	–
Ventr.	9.1	16.0	4.2	–	–	–	–	–	–
Ventr.	9.2	15.4	–	2.3	3.5	–	–	–	–
Ventr.	9.3	15.0	4.7	2.6	3.3	–	–	–	–
Ventr.	9.5	15.4	–	2.9	3.5	–	–	–	–
Ventr.	9.6	14.9	4.8	2.4	3.8	–	–	–	–
Ventr.	9.8	17.7	3.7	2.5	4.5	–	–	–	–
Ventr.	9.8	12.5	4.6	2.8	3.4	–	–	–	–
Ventr.	9.8	19.5	5.3	–	–	–	–	–	–
Ventr.	9.9	11.8	5.3	3.4	3.4	–	–	–	–
Ventr.	9.9	13.6	3.8	3.3	4.0	–	–	–	–
Ventr.	10.4	15.2	3.1	2.2	3.2	–	–	–	–
Ventr.	10.5	15.7	–	1.6	4.7	–	–	–	–
Ventr.	10.7	18.2	4.6	2.9	4.8	–	–	–	–
Ventr.	10.7	13.5	4.8	2.6	3.3	–	–	–	–
Ventr.	–	–	–	3.8	3.4	–	–	–	–

Leptestiina derfelensis (Jones, 1928)

Valve	L	W	D	L_msc	W_msc	L_ca	W_ca	L_be	W_be
Dors.	3.5	8.2	–	–	–	0.6	1.7	2.2	3.8
Dors.	3.9	8.6	–	–	–	0.6	2.3	2.5	4.1
Dors.	4.1	8.4	–	–	–	0.5	1.7	2.5	3.7
Dors.	4.6	7.3	0.7	–	–	0.6	1.7	2.7	4.5
Dors.	4.7	7.6	–	–	–	0.5	1.6	2.9	3.5
Dors.	6.3	12.8	1.5	–	–	–	–	–	–
Dors.	–	8.6	–	–	–	0.7	1.4	2.2	3.8
Dors.	–	8.1	–	–	–	–	1.7	2.3	3.6
Ventr.	3.3	7.6	0.9	0.7	1.6	–	–	–	–
Ventr.	3.5	6.0	–	–	–	–	–	–	–
Ventr.	3.5	9.1	–	0.7	1.8	–	–	–	–
Ventr.	3.6	8.0	0.8	0.7	1.5	–	–	–	–
Ventr.	3.7	8.4	1.0	0.7	1.5	–	–	–	–
Ventr.	3.9	8.2	1.7	0.7	1.6	–	–	–	–
Ventr.	3.9	7.8	1.7	0.9	1.5	–	–	–	–
Ventr.	3.9	8.4	1.4	0.7	1.5	–	–	–	–
Ventr.	4.0	8.6	1.1	1.1	1.6	–	–	–	–
Ventr.	4.1	8.2	1.0	0.7	1.5	–	–	–	–
Ventr.	4.2	7.5	1.3	0.9	1.2	–	–	–	–
Ventr.	4.3	9.4	1.3	0.8	2.0	–	–	–	–
Ventr.	4.4	7.7	1.2	0.8	1.5	–	–	–	–
Ventr.	4.4	6.8	1.9	0.7	1.3	–	–	–	–
Ventr.	4.5	7.1	1.2	0.8	1.3	–	–	–	–
Ventr.	4.5	9.7	1.5	1.1	1.6	–	–	–	–
Ventr.	4.6	8.0	–	–	–	–	–	–	–
Ventr.	4.6	7.9	1.6	1.0	1.4	–	–	–	–
Ventr.	4.7	7.8	1.7	1.3	1.7	–	–	–	–
Ventr.	4.8	9.6	1.8	0.9	1.5	–	–	–	–
Ventr.	4.8	9.5	1.3	0.9	1.7	–	–	–	–
Ventr.	5.0	9.8	1.5	0.9	1.9	–	–	–	–
Ventr.	5.6	9.9	1.6	1.2	1.7	–	–	–	–
Ventr.	5.8	10.7	1.8	1.4	2.1	–	–	–	–
Ventr.	6.6	12.8	2.7	1.4	2.2	–	–	–	–
Ventr.	6.8	10.5	2.7	1.1	2.4	–	–	–	–
Ventr.	7.7	11.2	2.6	2.2	2.5	–	–	–	–
Ventr.	9.3	13.1	2.6	1.8	2.6	–	–	–	–

Paurorthis aff. *Paurorthis parva* (Pander, 1830)

Valve	L	W	W_hi	D	L_msc	W_msc	L_br	W_br	L_int
Dors.	5.5	8.3	5.5	–	2.0	3.3	1.1	2.6	–
Dors.	6.1	7.2	–	–	–	–	1.4	2.7	–
Dors.	6.1	8.7	–	–	2.5	3.5	1.3	3.4	–
Dors.	6.4	9.0	–	–	–	–	1.4	3.5	–
Dors.	7.8	9.5	–	–	3.5	4.5	1.9	4.0	–
Dors.	8.0	9.5	–	2.2	–	–	–	–	–
Dors.	8.5	9.5	–	–	–	–	1.8	3.6	–
Dors.	10.5	11.7	–	2.4	–	–	–	–	–
Ventr.	6.2	6.7	4.2	–	2.2	1.3	–	–	1.0
Ventr.	7.8	7.2	–	–	3.0	1.6	–	–	1.2
Ventr.	8.0	8.1	5.8	–	3.1	1.8	–	–	–
Ventr.	8.1	7.5	6.6	–	3.2	1.7	–	–	1.5
Ventr.	8.2	9.3	6.0	–	2.7	1.7	–	–	–
Ventr.	8.5	9.5	–	3.1	–	–	–	–	–
Ventr.	9.8	8.6	–	–	3.3	1.6	–	–	1.5

Platystrophia tramorensis n. sp.

Valve	L	W	W_hi	D	L_msc	W_msc	W_fold	W_sulc
Dors.	4.4	–	–	–	–	–	–	–
Dors.	5.4	12.5	9.5	–	–	–	3.2	–
Dors.	5.7	10.0	–	2.7	1.8	2.2	3.0	–
Dors.	6.3	12.5	–	–	–	–	3.0	–
Dors.	6.4	8.7	7.2	2.0	–	–	1.6	–
Dors.	6.5	10.5	9.3	–	–	–	2.3	–
Dors.	7.0	9.5	8.5	3.2	2.8	2.7	2.7	–
Dors.	7.2	13.0	–	–	–	–	2.5	–
Dors.	7.3	10.1	8.5	–	–	–	2.6	–
Dors.	7.4	11.2	10.0	4.0	2.6	3.2	2.7	–
Dors.	7.8	12.2	10.3	–	–	–	2.2	–
Dors.	8.0	10.8	9.5	–	–	–	2.2	–
Dors.	8.1	10.2	10.2	3.8	3.2	3.7	2.0	–
Dors.	8.7	15.5	14.0	3.5	–	–	3.4	–
Dors.	8.8	12.5	10.8	3.6	2.5	3.7	–	–
Dors.	8.9	15.8	15.6	5.4	4.7	4.8	2.0	–
Dors.	9.0	13.3	–	–	–	–	3.0	–
Dors.	9.0	16.5	14.5	–	–	–	3.0	–
Dors.	9.2	11.5	9.0	–	–	–	2.4	–
Dors.	9.2	16.2	13.8	5.8	4.2	4.8	2.6	–
Dors.	9.5	12.3	12.0	4.3	3.5	4.2	2.4	–
Dors.	9.5	16.0	14.5	4.4	–	–	3.2	–
Dors.	10.3	11.2	–	3.2	3.0	2.7	–	–
Dors.	10.4	15.4	14.8	4.6	3.3	5.0	2.5	–
Dors.	–	10.0	9.8	3.7	3.0	3.0	2.5	–
Dors.	–	10.5	9.5	>4.0	–	–	–	–
Dors.	–	–	12.0	–	–	–	–	–
Ventr.	5.8	7.2	6.5	1.7	–	–	–	–
Ventr.	5.8	10.2	–	–	–	–	–	–
Ventr.	6.2	7.4	5.8	1.7	2.5	1.3	–	–
Ventr.	6.2	7.5	6.3	2.2	2.2	1.1	–	–
Ventr.	6.2	7.5	6.0	2.7	2.2	1.0	–	2.2
Ventr.	6.4	8.7	7.2	–	–	–	–	2.7
Ventr.	6.5	8.3	7.2	2.0	3.4	1.7	–	2.0
Ventr.	7.0	9.2	7.2	–	–	–	–	1.7
Ventr.	7.3	10.1	8.5	–	–	–	–	1.8
Ventr.	7.3	12.2	–	3.3	2.5	1.3	–	3.0
Ventr.	7.8	12.2	10.3	–	–	–	–	2.1
Ventr.	8.0	10.8	9.5	–	–	–	–	1.7
Ventr.	8.0	15.0	11.5	2.8	2.7	1.9	–	2.4
Ventr.	8.2	11.5	11.0	3.5	3.6	2.0	–	–
Ventr.	8.6	12.3	–	–	–	–	–	2.5
Ventr.	8.6	15.0	11.0	2.4	3.4	1.6	–	2.5
Ventr.	9.0	16.5	14.5	–	–	–	–	2.4
Ventr.	9.2	11.5	9.0	–	–	–	–	1.6
Ventr.	9.2	16.2	13.8	3.7	4.3	2.5	–	2.5
Ventr.	9.5	13.5	–	2.3	–	–	–	3.5
Ventr.	9.8	11.8	10.2	2.3	4.1	1.6	–	–
Ventr.	10.2	–	9.4	3.2	4.2	1.7	–	2.5
Ventr.	10.3	16.5	–	2.3	3.2	2.0	–	–
Ventr.	10.5	13.0	–	–	–	–	–	3.2
Ventr.	10.5	14.5	12.0	4.5	3.8	1.3	–	–
Ventr.	12.0	17.5	–	3.6	4.7	1.7	–	2.5
Ventr.	–	10.5	9.5	4.0	–	–	–	2.2
Ventr.	–	–	–	–	–	–	–	2.6

Platystrophia aff. *Platystrophia sublimis* Öpik, 1930

Valve	L	W	W_hi	D	L_msc	W_msc	W_fold	W_sulc
Dors.	3.2	6.2	–	–	–	–	–	–
Dors.	3.5	5.8	–	–	–	–	1.7	–
Dors.	4.5	5.6	–	–	–	–	1.2	–
Dors.	6.0	9.2	–	–	–	–	–	–
Dors.	6.3	9.2	8.2	2.5	3.6	4.1	1.7	–
Dors.	7.1	7.8	7.3	–	1.6	1.8	2.0	–
Dors.	7.5	9.8	9.3	2.9	–	–	–	–
Dors.	8.6	13.0	–	–	–	–	2.5	–
Dors.	–	–	8.5	3.0	–	–	–	–
Dors.	–	19.2	19.0	–	4.0	4.2	4.3	–
Ventr.	3.3	5.8	–	–	–	–	–	–
Ventr.	4.5	5.8	–	–	–	–	–	1.8
Ventr.	4.7	>5.0	–	–	–	–	–	–
Ventr.	6.4	10.0	9.5	2.0	1.7	1.2	–	–
Ventr.	7.1	9.3	–	–	–	–	–	–
Ventr.	7.5	9.8	9.3	2.6	–	–	–	–
Ventr.	8.4	13.0	–	–	3.5	1.5	–	–
Ventr.	9.0	13.3	–	–	3.3	1.3	–	2.1
Ventr.	–	–	–	–	–	–	–	–
Ventr.	–	–	–	–	–	–	–	–
Ventr.	–	–	–	–	–	–	–	1.6
Ventr.	–	–	–	–	–	–	–	–
Ventr.	–	–	–	2.0	–	–	–	–

Productorthis sp. A

Valve	L	W	W_hi	D	L_msc	W_msc	L_br	W_br
Dors.	13.1	17.6	15.2	5.2	7.0	7.0	2.6	6.0

Salopia gracilis Williams *in* Whittington & Williams, 1955

Valve	L	W	D	L_msc	W_msc	L_ms	L_ca	L_int
Dors.	4.2	7.5	–	1.5	1.8	2.3	–	–
Dors.	4.3	5.6	–	1.8	2.1	3.6	0.8	–
Dors.	4.7	6.1	–	1.7	1.8	3.1	0.9	–
Dors.	5.3	7.6	1.3	2.0	3.1	3.4	0.9	–
Dors.	5.3	7.8	–	–	–	–	–	–
Dors.	6.1	7.3	–	2.7	2.5	3.8	1.1	–
Dors.	6.6	8.3	–	2.6	3.3	4.2	1.1	–
Dors.	6.8	9.2	–	–	–	5.5	2.2	–
Dors.	7.6	9.5	–	–	–	–	–	–
Dors.	7.7	8.9	–	3.5	3.3	5.3	–	–
Dors.	7.7	10.3	–	–	–	5.7	2.0	–
Dors.	7.8	8.2	–	–	–	–	–	–
Dors.	8.3	9.3	1.0	3.7	4.1	6.7	1.7	–
Dors.	8.4	9.1	–	3.0	3.7	6.2	2.3	–
Dors.	9.6	11.2	–	–	–	6.5	–	–
Dors.	9.9	12.4	–	–	–	6.2	–	–
Dors.	11.5	14.2	–	4.3	6.3	7.6	2.4	–
Ventr.	6.5	–	–	–	–	–	–	1.5
Ventr.	7.3	10.2	–	2.9	1.7	–	–	1.7
Ventr.	7.7	11.1	–	2.6	2.3	–	–	–

(continued)

Salopia gracilis Williams *in* Whittington & Williams, 1955. (continued)

Valve	L	W	D	L_msc	W_msc	L_ms	L_ca	L_int
Ventr.	7.8	10.0	–	3.1	2.1	–	–	–
Ventr.	8.3	8.3	2.7	3.2	1.7	–	–	–
Ventr.	9.0	10.1	–	3.7	2	–	–	–
Ventr.	10.4	8.5	–	4.3	2.5	–	–	3.0
Ventr.	11.8	13.9	–	4.5	3.2	–	–	–
Ventr.	12.2	11.2	–	4.8	3.1	–	–	–

Sowerbyella (*Sowerbyella*) *antiqua* Jones, 1928 (Table 1)

Valve	L	W	L_msc	W_msc	Bema	C_be	Dir (°)
Dors.	0.9	2.0	–	–	–	–	318
Dors.	2.0	–	–	–	No bema, only septa	1	–
Dors.	2.1	4.4	–	–	No bema, only septa	1	225
Dors.	2.2	6.2	–	–	Moderately raised	2	250
Dors.	2.4	7.0	–	–	No bema, only septa	1	–
Dors.	2.5	6.4	–	–	Raised	2	28
Dors.	2.5	–	–	–	Raised	2	313
Dors.	2.9	5.8	–	–	–	–	23
Dors.	2.9	6.0	–	–	No bema, only septa	1	59
Dors.	3.0	6.1	–	–	No bema, only septa	1	47
Dors.	3.0	6.4	–	–	Moderately raised	2	90
Dors.	3.0	8.1	–	–	No bema, only septa	1	193
Dors.	3.1	6.1	–	–	Raised	2	–
Dors.	3.1	6.5	–	–	No bema, only septa	1	140
Dors.	3.1	6.6	–	–	Raised	2	217
Dors.	3.2	6.2	–	–	No bema, only septa	1	190
Dors.	3.2	7.4	–	–	Raised	2	–
Dors.	3.5	6.2	–	–	–	–	70
Dors.	3.7	5.9	–	–	–	–	342
Dors.	3.8	6.1	–	–	–	2	–
Dors.	3.8	7.0	–	–	Moderately raised	2	251
Dors.	3.9	6.8	–	–	Raised	2	54
Dors.	4.0	6.0	–	–	–	–	23
Dors.	4.0	10.5	–	–	Moderately raised	2	204
Dors.	4.2	6.4	–	–	Raised	2	–
Dors.	4.2	7.2	–	–	Moderately raised	2	298
Dors.	4.4	7.7	–	–	Raised	2	272
Dors.	4.4	9.2	–	–	Raised	2	318
Dors.	4.5	6.0	–	–	Raised	2	131
Dors.	4.5	7.1	–	–	Raised	2	248
Dors.	4.5	10.1	–	–	Moderately raised	2	236
Dors.	4.8	7.0	–	–	Raised	2	2
Dors.	6.1	–	–	–	–	2	–
Dors.	–	4.4	–	–	Raised	2	140
Dors.	–	9.0	–	–	Raised	2	200
Ventr.	1.3	3.8	–	–	–	–	204
Ventr.	2.0	4.2	–	–	–	–	264
Ventr.	2.5	5.1	–	–	–	–	74
Ventr.	2.9	7.2	–	–	–	–	24
Ventr.	3.0	6.0	0.9	1.0	–	–	260
Ventr.	3.0	7.8	0.9	1.7	–	–	67
Ventr.	3.3	6.4	–	–	–	–	217
Ventr.	3.3	7.6	–	–	–	–	324
Ventr.	3.3	7.8	–	–	–	–	183
Ventr.	3.4	6.8	–	–	–	–	219
Ventr.	3.4	8.0	1.1	1.4	–	–	23
Ventr.	3.5	10.0	–	–	–	–	47
Ventr.	3.6	5.9	1.1	1.5	–	–	183
Ventr.	3.6	7.0	–	–	–	–	25
Ventr.	3.7	8.2	–	–	–	–	195
Ventr.	3.8	9.7	–	–	–	–	20

(continued)

Sowerbyella (*Sowerbyella*) *antiqua* Jones, 1928 (Table 1). (continued)

Valve	L	W	L_msc	W_msc	Bema	C_be	Dir (°)
Ventr.	3.9	6.1	–	–	–	–	–
Ventr.	3.9	6.8	1.1	1.3	–	–	180
Ventr.	3.9	8.0	–	–	–	–	–
Ventr.	3.9	9.9	1.0	1.8	–	–	220
Ventr.	4.0	5.7	–	–	–	–	50
Ventr.	4.0	6.8	0.7	1.2	–	–	–
Ventr.	4.0	8.0	1.0	1.8	–	–	80
Ventr.	4.0	9.8	–	–	–	–	34
Ventr.	4.0	9.8	0.9	1.3	–	–	41
Ventr.	4.1	7.6	1.8	1.9	–	–	356
Ventr.	4.1	10.4	–	–	–	–	19
Ventr.	4.2	6.0	–	–	–	–	296
Ventr.	4.2	8.1	–	–	–	–	216
Ventr.	4.2	8.4	–	–	–	–	–
Ventr.	4.3	7.6	–	–	–	–	104
Ventr.	4.7	7.0	1.0	1.8	–	–	291
Ventr.	4.7	8.2	–	–	–	–	160
Ventr.	4.8	8.2	1.0	1.5	–	–	162
Ventr.	5.0	6.8	–	–	–	–	253
Ventr.	5.0	8.9	0.9	1.2	–	–	104
Ventr.	5.0	–	–	–	–	–	–
Ventr.	–	–	–	–	–	–	–

Sowerbyella (*Sowerbyella*) *antiqua* Jones, 1928 (Table 2)

Valve	L	D	L_ms	L_ca	W_ca
Dors.	2.2	–	1.4	0.4	0.7
Dors.	2.4	–	1.5	0.2	0.8
Dors.	2.9	–	1.5	0.3	0.9
Dors.	2.9	–	1.6	0.3	1.1
Dors.	2.9	–	1.6	0.3	0.9
Dors.	3.0	–	1.7	0.3	1.2
Dors.	3.0	–	2.0	0.3	1.2
Dors.	3.2	–	2.2	0.4	1.5
Dors.	3.3	–	1.8	0.3	1.2
Dors.	3.3	–	1.8	0.3	1.3
Dors.	3.3	–	2.2	0.4	1.3
Dors.	3.4	–	1.5	0.3	1.3
Dors.	3.4	–	1.7	0.4	1.3
Dors.	3.4	–	1.8	0.4	1.0
Dors.	3.5	–	1.8	0.3	1.0
Dors.	3.6	–	2.0	0.3	1.4
Dors.	3.8	–	1.4	0.4	1.4
Dors.	3.8	–	1.7	0.4	1.5
Dors.	3.8	–	1.8	0.4	1.4
Dors.	3.8	–	1.9	0.4	1.6
Dors.	3.8	–	2.2	0.4	1.5
Dors.	4.0	–	1.7	0.4	1.7
Dors.	4.0	–	1.7	0.4	1.3
Dors.	4.1	–	1.9	0.4	1.3
Dors.	4.3	–	1.3	0.4	1.4
Dors.	4.3	–	2.3	0.4	1.4
Dors.	4.8	–	1.9	0.4	1.7
Dors.	4.9	–	2.2	0.5	1.6
Dors.	5.3	–	2.6	0.5	1.7
Dors.	5.3	–	2.8	0.5	1.8
Ventr.	2.0	0.3	–	–	–
Ventr.	2.1	0.5	–	–	–
Ventr.	3.6	1.1	–	–	–
Ventr.	3.6	0.7	–	–	–
Ventr.	3.6	0.6	–	–	–
Ventr.	3.6	0.7	–	–	–

(continued)

Sowerbyella (*Sowerbyella*) *antiqua* Jones, 1928 (Table 2). (continued)

Valve	L	D	L_ms	L_ca	W_ca
Ventr.	3.7	0.5	–	–	–
Ventr.	3.7	0.9	–	–	–
Ventr.	3.7	1.4	–	–	–
Ventr.	3.9	1.1	–	–	–
Ventr.	3.9	0.9	–	–	–
Ventr.	4.1	0.9	–	–	–
Ventr.	4.3	0.8	–	–	–
Ventr.	4.5	1.2	–	–	–
Ventr.	4.6	0.6	–	–	–
Ventr.	4.7	1.1	–	–	–
Ventr.	4.7	1.0	–	–	–
Ventr.	4.8	1.1	–	–	–
Ventr.	4.9	1.3	–	–	–
Ventr.	5.5	1.4	–	–	–

Sulevorthis aff. *Sulevorthis blountensis* (Cooper, 1956)

Valve	L	W	D	L_msc	W_msc	L_br	W_br
Dors.	3.1	4.0	–	–	–	–	–
Dors.	3.4	4.5	–	–	–	0.6	1.4
Dors.	3.7	5.0	–	–	–	0.8	1.6
Dors.	3.7	6.0	–	–	–	–	–
Dors.	5.5	7.2	–	–	–	–	–
Dors.	6.1	9.8	–	–	–	1.1	3.0
Dors.	6.2	7.8	–	–	–	1.2	2.5
Dors.	6.2	8.5	–	–	–	–	–
Dors.	7.0	10.4	–	–	–	1.3	3.3
Dors.	7.1	8.6	–	–	–	1.2	2.8
Ventr.	1.9	2.9	–	–	–	–	–
Ventr.	3.0	3.5	–	0.8	0.7	–	–
Ventr.	3.2	4.2	1.6	–	–	–	–
Ventr.	3.6	4.5	–	1.3	0.8	–	–
Ventr.	3.7	–	–	1.2	–	–	–
Ventr.	4.3	5.6	–	–	–	–	–
Ventr.	4.5	6.0	–	–	–	–	–
Ventr.	4.5	6.2	–	1.5	1.1	–	–
Ventr.	5.1	6.6	–	–	–	–	–
Ventr.	6.5	7.8	–	–	–	–	–
Ventr.	6.5	9.5	–	–	–	–	–
Ventr.	7.5	10.4	–	–	–	–	–

Tetraphalerella? sp. A

Valve	L	W	L_msc	W_msc	L_gen	A_dp
Dors.	26.5	34.0	10.5	12.4	23.0	135°

Valcourea confinis (Salter, 1849)

Valve	L	W	D	L_msc	W_msc	L_ms	L_br	W_br
Dors.	11.9	18.7	2.4	–	–	4.3	1.7	3.2
Dors.	13.9	19.0	–	–	–	6.0	2.7	4.1
Dors.	15.0	21.2	2.7	–	–	–	–	–
Ventr.	5.3	9.0	–	–	–	–	–	–
Ventr.	9.8	16.4	–	–	–	–	–	–
Ventr.	12.5	24.3	–	5.0	7.3	–	–	–